P9-AOS-325

THE MISSING MOMENT

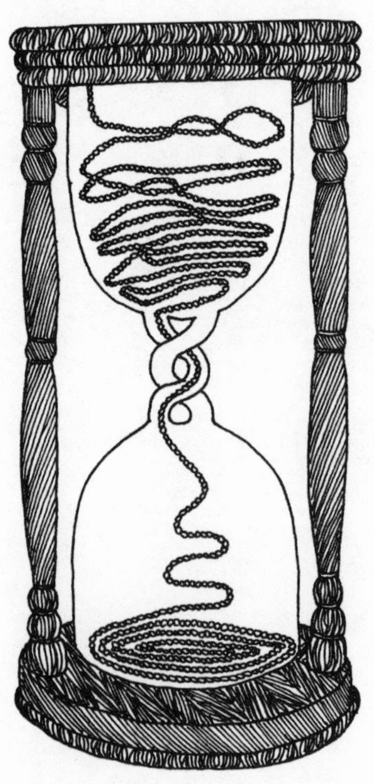

The Missing Moment

HOW THE UNCONSCIOUS SHAPES MODERN SCIENCE

Robert Pollack

Houghton Mifflin Company

BOSTON NEW YORK

1999

Library of Congress Cataloging-in-Publication Data

Pollack, Robert, date.
The missing moment : how the unconscious shapes modern science / Robert Pollack.
p. cm.
Includes index.
ISBN 0-395-70985-7
1. Medicine and psychology. 2. Medicine — Philosophy. I. Title.
R726.5.P63 1999 610'.1—dc21
99-26241 CIP

Frontispiece by Amy Pollack
Book design by Anne Chalmers
Typeface: Sabon; Linotype-Hell Didot

Printed in the United States of America

QUM 10 9 8 7 6 5 4 3 2 1

To Molly Pollack, 1913–1996

And it is time, strength, tone, light, life and love—
And even substance lapsing unsubstantial;
The universal cataract of death
That spends to nothingness—and unresisted,
Save by some strange resistance in itself,
Not just a swerving, but a throwing back,
As if regret were in it and were sacred.
It has this throwing backward on itself
So that the fall of most of it is always
Raising a little, sending up a little.
Our life runs down in sending up the clock.
The brook runs down in sending up our life.
The sun runs down in sending up the brook.
And there is something sending up the sun.
It is this backward motion toward the source,
Against the stream, that most we see ourselves in,
The tribute of the current to the source.
It is in this from nature we are from.
It is most us.

—Robert Frost, *Westward Running Brook*

ACKNOWLEDGMENTS

THIS BOOK is as much about the healing power of love as it is is about the future of medicine. The support and encouragement of my family—my wife, Amy, our daughter, Marya, her husband, Mark Lehrman, and my brother, Barry—were an act of love and a blessing to me.

With the support of the Ford Foundation, the Richard Scheuer Foundation, the Sloan Foundation, the Milstein Family Foundation, the Lucius Littauer Foundation, and the Office of the Vice Provost of Columbia University, I was able to complete this book during a sabbatical break from my regular teaching duties at Columbia. This book also benefited from my extended Upper West Side family: friends and colleagues from Columbia—in particular, members of the Department of Biological Sciences, the Center for Psychoanalytic Training and Research, and the University Seminar on Human Diversity—from Congregation B'nai Jeshurun, and from the Jewish Theological Seminary of America gave various drafts their close attention.

Those kind colleagues whose business it is to make books come out well carried out that business with great flair and kindness, and for both I am most grateful. My editor at Houghton Mifflin, Harry Foster, was my most perceptive and gentlest critic, and Luise Erdmann read the manuscript with unique care; Jean Naggar and Anne Engel of the J. V. Naggar Literary Agency were unfailing in their support and encouragement.

The advice of Judith Berman, Kate Brauman, Beth Brodsky, Marcello Bronstein, Audrey Chapman, Jonathan Cole, Michael Crow, Margy-Ruth and Perry Davis, Neil Gillman, Robert Glick, Ari Hakimi, Jonathan House, Eric Kandel, Darcy Kelley, Henry Kennedy, Alex Levay, Rolando Matalon, Kathleen McDermott, Don Melnick, David Olds, Robert O'Meally, Lynn Paltrow, Ellen Peyser, Barrie Raik, William Sage, Joanna Samuels, Amir Shaviv, Josh Schulman, Leonard Sharzer, Dovid Silber, Devorah Steinmetz, Adin Steinsaltz, Joe Thornton, Jan Urbach, and Miriam Warshaviak made this a better book.

Despite the best efforts of all family, friends, and colleagues, mistakes of omission are certain to have occurred, and misstatements of fact may well have slipped by as well. For both, I take full responsibility. A century ago Freud wrote of his hopes for *The Interpretation of Dreams:* "It is [my] earnest wish that the book age rapidly—that what was once new in it may become generally accepted, and that what is imperfect in it may be replaced by something better." I wish the same for this book.

New York City and Chelsea, Vermont
February 1999

Contents

THE MISSING MOMENT

Introduction

Contrary to what I had believed, the process of experimental science does not consist in explaining the unknown by the known, as in certain mathematical proofs. It aims, on the contrary, to give an account of what is observed by the properties of what is imagined.

— Francis Jacob, *The Statue Within*

"A sigh for ever so many a breath,
And for ever so many a sigh a death.
That's what I always tell my wife
Is the multiplication table of life."
The saying may be ever so true;
But it's just the kind of a thing that you
Nor I, nor nobody else may say,
Unless our purpose is doing harm,
and then I know of no better way
To close a road, abandon a farm,
Reduce the births of the human race,
And bring back nature in people's place.

— Robert Frost, *The Times Table*

THIS BOOK IS about the difference between scientific knowledge and scientific wisdom. Knowledge is not wisdom. King Solomon knew the difference, and when he had to choose between them, he picked wisdom. Two prostitutes came to him with a difficult case. They had both given birth within the past three days and the mothers and newborns were alone together the night one infant died. Each claimed the remaining child was hers. Kings I, Chapter 3, describes Solomon's strategy for deciding between them:

> The king said, "One says, 'This is my son, the live one, and the dead one is yours'; and the other says, 'No, the dead boy is yours, mine is the live one.'" So the king gave the order, "Fetch me a sword." A sword was brought before the king, and the king said, "Cut the live child in two, and give half to one and half to the other."
>
> But the woman whose son was the live one pleaded with the king, for she was overcome with compassion for her son. "Please, my lord," she cried, "give her the live child, only don't kill it!" The other insisted, "It shall be neither yours nor mine; cut it in two!" The king spoke up. "Give the live child to her," he said, "and do not put it to death; she is its mother."[1]

Solomon had no way of knowing which woman was the biological mother, but he took the absence of such knowledge as an opportunity for wisdom and gave the baby to the woman who displayed the most compassion for it. Now imagine that Solomon's court was being held today, and the woman who called out for the child to be divided in two, who felt so strongly about losing it that she would rather see it dead than in the other's hands, said, "Please, my lord, I appeal to you for a DNA test." What would Solomon do if the test showed that she was, in fact, the mother of the live child? Would this knowledge lead him to change his decision, or would his wisdom still compel him to let it stand? The text makes it plain that Solomon's wisdom lay in deciding where the child's interests lay, and that its interests neither depended on nor could be served by any manner of scientific evidence. What was wise for Solomon to do would be wise for us as well: we should expect medicine and science to place human needs ahead of other considerations and to be sensitive to the facts of life and death that unite us all.

Holding on to the wisdom of Solomon in the age of DNA will not be easy. The live child did not need to know its mother's DNA, it needed to be cared for. Its life would not have been in any way improved by genomic knowledge. Many of us will face a dilemma like Solomon's in years to come. Each year molecular medicine tells more of us about the inherited conditions from which we will fall ill or even die without being able to cure, prevent, or even ameliorate many of these conditions. We are caught in a time when knowledge falls between ignorance and wisdom. To use human genes as tools for the creation of new treatments for life-threatening diseases would be the heart of wisdom; to use them simply to foretell a person's future is to burden ourselves with an ever-heavier load of useless, and even damaging, knowledge.

Why is there not more wisdom in the application of scientific discoveries to the lives of sick and suffering people? Why is there not more wisdom in plans for the future of both medical science and medical practice? These questions are not ordinarily asked by scientists and doctors. This reticence is, I think, a major clue, a door to go through in order to understand why today's medicine seems so full of knowledge yet so far from wise. There are many other doors for others to open — sociology, anthropology, economics, politics, and religion come to mind — but I am myself a scientist, and this is first and foremost a book about how science works and how my branch of science might work to the greater benefit of us all. I have taken a new look at some important results of the very science I am most concerned about and found that a lack of wisdom in science may be the natural but avoidable consequence of a gap between the way the senses and the mind experience time, on the one hand, and the way science works, on the other.

•

Time's passing is the frame around any scientific model's testability and reproducibility; it is the passive but stable background against which all measured change, in any aspect of our bodies or minds, takes place. The unchallenged stability of the laboratory clock gives it a privileged place in today's science; it is the premier tool of scientific measurement, the single stable fixture in an otherwise ever-changing universe. The universe itself — and time and space as well — began from a single point some ten to twenty billion years ago. The beginning instant has left strong evidence of its occurrence, but dating it more precisely than that has been a little like looking for a signature on water. Ever since that instant, as the universe has expanded and cooled, all but one aspect of the natural world has been changing. Today the universe is a thin, lumpy, expanding cosmic soup, its ingredients still not quite stable, its parts still not quite held together by the forces through which they interact with one another. Only the speed of a unit of light — a photon traveling through the vacuum of empty space — has been, is, and so far as we can tell always will be the same: 299,792,458 meters each second.

Though this number is based on various assays, it is also a singular icon of the scientific world. It is a universal law, a statement of unique, universal consequence. It is an absolute limit: nothing can go faster than light. The universe's physical consistency from one edge to another — and so also our ability to understand it through the tools of science — is wholly dependent on the stability of the speed of light through empty space. We choose to measure the passage of time by splitting it into bits and counting them — seconds, years, generations — but the essence of time's passage is its smoothness.

Einstein saw that time reveals itself to be not only something measured but also a dimension inextricably interwoven with the three dimensions of spatial reality. As the mathematician Hermann Minkowsky put it about a century ago, "Nobody has ever noticed a place except at a time, or a time except at a place. Henceforth; space by itself, and time by itself, are doomed to fade away into mere shadows, and only a kind of union of the two will preserve an independent reality." No physical object can be perfectly stable; everything becomes different at different times, simply because its place in time has changed.[2]

Time's passage outside ourselves in the objective, measurable world may not be deflected from its inexorable course, but time's passage inside a person's head is something else entirely. Einstein's grand conception of the relativistic universe hints at this more flexible, malleable notion of time for conscious minds like ours. Our perception of the outside world can be slowed, stopped, reversed, or sped up by the ticking of a number of internal clocks. Two very old clocks build the body from a single cell, a third drives the mechanisms of perception, another keeps us in synchrony with day and night, and still others create a conscious sense of the world and link it to unconscious memory. The inner times created by these clocks are multiple and complex, coming together only once at the moment death brings them all to a stop.

These clocks do not affect objective time, but they profoundly alter the time we sense inside ourselves. The conscious experience of time is complicated, flexible, and even multiple. It passes quickly when we concentrate, slowly when we are bored, not at all when we sleep, and it reverses itself with every memory. As we experience the world through our senses, think about our experiences, put some of our thoughts and experiences in memory, or recall them at some later time, these clocks ground the experience of the moment in the past.

Science is a creation of the human mind that appears to live entirely in the present and to be able to use the passage of time as a tool, but it too is the product of brains that create consciousness by binding together perceptions of the moment with feelings and memories from the past. For the most part, science has been able to proceed along its grand agenda of understanding the universe by imagination and experiment without having to deal with the differences between external and inner time. In the past few years, though, experiments performed in smooth scientific time have shown that the times these inner clocks keep are as uncoupled from the conscious sense of time's passage, which they themselves create, as they are from objective time.

The inner clocks of the body and mind were discovered by many scientists working in many disciplines at different times. So it has taken a while to notice that all together these recent discoveries explain why the brains and minds of everyone — even scientists — are designed in such a way that they will always be driven — consciously or unconsciously — by the past. All these conscious and unconscious clocks that tick away internal time's passage are products of the past, of the evolution of our species. By exploring the way natural selection has combined sensations and memories in the brain to give each person an inner voice, we can see how scientists themselves may be constrained by the very mechanisms they have recently discovered. They have found out, for instance, why the mind cannot function without constant reference to the past. Some aspects of the past are retained as conscious memories, others can be recalled when needed, and still others — especially those fraught with pain or fear — lie embedded in memory but inaccessible to conscious recall. Unconscious — repressed — memories are an essential part of all conscious mental functions, including the ones we call science and medicine. Of these fears, the fear of total loss of oneself by death is surely the oldest, deepest, and most intense, and therefore it is the one most likely to be kept from consciousness.

Yet at the same time science has found that individual mortality has been a fact of life since the living world began: neither the best nor the worst, but simply the only strategy for life's ongoing survival. Life's dependence on DNA for its overall survival is the key to life's dependence on individual mortality: the replication of DNA provides continuity from generation to generation despite it, as mutational errors during DNA replication provide the heritable variation from individual to individual that natural selection must have. For DNA to provide natural selection with sufficient fresh, transient material for evolution to occur, DNA-based life-forms must always have had limited individual lifetimes. The result of the union of DNA's stability with any one life's evanescence has been a four-billion-year-old, still-diversifying tree of mortal life-forms including ourselves, each connected to all the others by common descent. Just as the seeds of a tree hang from branches made largely of dead wood, creatures alive today are here only because of the death of ancestral species. As Darwin wrote:

> As buds give rise by growth to fresh buds, and these, if vigorous, branch out and overtop on all sides many a feebler branch, so by generation I believe it has been with the great Tree of Life, which fills with its dead and broken branches the crust of the earth, and covers the surface with its ever-branching and beautiful ramifications.

The passage of time in any living thing is thus linked to the certainty of death. To know that time is passing through me and that I am different at every instant because of its passage is to know — and to face the knowledge — that I must die. This consequence of mortality creates a little-appreciated but critical difference in the way time is experienced by any living thing and the way it has been adopted by science as a tool. Even in the shortest instant of conscious thought we are older, one instant closer to the end of our inevitably limited time. Meanwhile, the isolated, pure time of science — external time — passes outside anything mortal, flowing on smoothly and wholly dissociated from the changes it may cause in any living thing. The dissociation of inner and outer times is the root cause of the difficulty scientists have in coming to terms with mortality. The passage of objective time is the premier instrument of science, and yet none of us — nor any living thing — can ever directly experience it. Mortality, the common fate that links us all, erects a wall of silence and blankness at the outer edges of our scientific understanding of the passage of time.

Like Mr. Scrooge's dreaming of a visit to his own grave, I once saw an early rounding off to my own life. In my case, as in his, it was not entirely a dream, nor could I continue unchanged afterward. In the spring of 1993 — as I was in Vermont completing my first book, *Signs of Life* — I came down with a bad cold that did not clear up. I would cough and feel tired, and I had developed a distressing capacity to produce vast amounts of phlegm. These symptoms did not slow me down enough to take me to a doctor, nor did I allow them to interfere with my summer plans. By September I had resigned myself to coughs and phlegm forever when my symptoms took a jump and I was knocked off my feet and into bed by a new, painful, racking cough and a high fever. I was miserable that night and the next day. I began to hallucinate: the fever split me mentally into a person who could not concentrate on anything and who saw whirling colors, who heard a small voice saying that this was bad but was unable to do much about it. My remaining powers of concentration barely allowed me to accomplish the tasks of sitting up to cough, and shuffling to the bathroom to relieve myself.

I was too sick to be scared, but my wife was afraid to let another night like that go by. She decided to consult our doctor in New York, who had me take an antibiotic drug, erythromycin. Within a day after my first dose, my fever had almost gone, and I was able to drive back to New York a few days later. X-rays confirmed the diagnosis of pneumonia. My lungs did not fully recover for another month, but I had come through. I owed my life to erythromycin, a drug that did not exist when I was born; fifty years ago I would have died. Although the antibiotic worked and left me physically whole, the pneumonia nevertheless changed me. Although mortality was of little or no interest to science, my bout with pneumonia had ended my complacent disinterest. The sight of my own death meant an end to my days of freedom to ignore mortality, and with them, my career as a laboratory scientist.

My reaction was so strong, perhaps, because of my particularly vulnerable age. I was in my mid-fifties, an age that bears a certain resemblance to a similar southern latitude. At latitudes near 50 degrees south, no landmasses impede the winds that swirl around the planet. Coming and going suddenly, they hit boats with startling ferocity; anyone who has tried to cross the south fifties knows how dangerous and difficult it is. Strong winds of a different sort hit all of us as we cross our own personal fifties. On the warm side, coming down from the equator of childhood, we measure our latitude by the time we've had on this earth; after the crossing, we confront the chilling question — how much time do we have left?

In my fifties, pneumonia took from me the one article of faith necessary to a career in biomedical research: that although science may not yet be able to explain everything worth knowing about, in due course it will be able to. Still in the full faith that this was so, I began to search for the place of mortality in the research agendas of my colleagues. I soon found that science had no useful model for dealing with mortality nor any apparent interest in developing one: death was simply not interesting. This struck me as odd.

Although I had just been saved by the right antibiotic — by knowledge more than wisdom — I knew that no state of scientific knowledge would save me forever, and I knew that each of my colleagues shared the same information. My new question about nature had become a pair of questions about science: why was the inevitability of death shoved aside when it was so clearly on everyone's mind, and what would science look like if it were able to admit this limitation, accept mortality, and focus on how best to make all our mortal lives last longer?

Not death itself, but the refusal to consciously contemplate death and the end of our personal inner time, continues to limit our capacity to use the discoveries of science to extend our lives to their fullest capacity. The difference between the inner time of a patient and the objective time of science may seem small to the scientist, but it is everything to the patient. Considering that doctors and scientists inevitably must become patients themselves sooner or later, it may seem paradoxical that so much of medical science ignores the differences between objective and personal time, acting as if inner times are less important than objective time. But there is a reason that individual mortal lives are certain to be less interesting to science and medicine than they should be. Biomedical scientists, who have built their models of our bodies and minds on the presumption that we exist only in the context of outside time, have found — through their work — a way to avoid confronting their own mortality.

Like everyone else, scientists think and feel with brains assembled and run by a host of different, even conflicting, inner times. The reality of science is thus a melding of a scientist's past with current sensations, perceptions, ideas, dreams, and needs. Scientific reality, like any other attempt to explain the world, lies, as Harold Bloom says, at the "rim of inner self"; it begins in the universe of our senses and ends at the moment of death. In between, throughout our lives, our conscious memories and unconscious selves bring our biological and personal histories to the rim. The capacity of the unconscious to collapse time — repressing early conscious experience but allowing its reappearance in later life — is the key to understanding why today's medical science fails to deal directly with the primary obligation of medicine: to extend the inner time a person has by preventing, curing, or ameliorating the diseases that bring on premature death.

Getting a scientific field to come to terms with fears its practitioners may unconsciously have experienced but cannot admit will be much like getting a person to see and acknowledge the ways that thoughts denied for years or decades may be contributing to a current difficulty. In the same way that uncovering repressed thoughts and fears can relieve a person of the burdens of the past and permit a freedom and range of self-expression in the present, so medical science is likely to become more interesting and more valuable to all of us once scientists acknowledge how the passage of different internal times affects their own bodies, their own minds, and therefore their own capacity to distinguish between knowledge and wisdom.

1

Sensation

All theory, my friend, is gray
But green is life's glad golden tree.
— Mephistopheles to his student Faust, in Goethe's *Faust*

Das Sein ist ewig; den Gesetze
Bewahren die lebend'gen Schätze,
Aus welchen sich das All geschmückt.
(Being is eternal; for there are laws to conserve the treasures of life, on which the Universe draws for its beauty.)
— Goethe, quoted in Erwin Schrödinger, *What Is Life?*

MAKING SENSE of the world is the job of science. In order for scientists to begin that work, they first must engage the world through their senses. The ways we sense the world — the way we see it or smell it, for instance — may seem consistent with some rational purpose, allowing us to grasp the world just as we would design a set of instruments to do so. But the opposite is the case, as neither our rational intentions nor our experimental models of the natural world inform the design of our senses. Though scientists and doctors may design their instruments of experimentation and diagnosis to observe the world with precision and accuracy, the senses they use to take in their results are designed and constantly rebuilt for other purposes by clocks inside the body. We are each home to many imperceptible clocks, some of which run as fast as a thousand ticks each second, others as slow as one tick per lifetime. These clocks share two attributes: they all operate without being noticed — although they can be found — and none of them uses time in quite the way any instrument of science does.

The oldest and slowest of the clocks that build the senses is one we share with all other forms of life: the clock of natural selection, which continually builds new forms of life from old. Its beat is the birth of a species; millions of years may go by between ticks. One of its products is the second inner clock, a rhythm of signals that turn genes on and off in the cells of a developing embryo. This rhythm creates the senses as it builds a human body from genetically identical cells descended from a single fertilized egg. Though it is not as ancient and pervasive as the first, this clock can be found inside all living things made of more than one cell. A third clock is more restricted; it ticks only in nerve cells, welding the nervous system together with impulses arriving less than a thousandth of a second apart. These three clocks are deeply embedded in the past, and thus the senses — the initial instruments of scientific observation that these clocks have assembled — are connected to the past as well. Our ways of knowing the world — our senses — may seem designed for our present needs, but because they are assembled by these three clocks, they were designed to meet the needs of species now long dead.

The clock that differentiates the many cells of a complex organism from one another works by turning different genes on or off in different cells. It starts up at fertilization, the very first instant in the development of an embryo. The plumbing involved in conception has been the object of obsessive attention since ancient times, but the actual events that mark the beginning of a new person from a single cell were clarified only in the past century or so. Aristotle taught — and many scientists through at least the first half of the nineteenth century accepted as fact — that a baby began when semen mingled with uterine secretions and caused them to coagulate; in this way the man provided the baby's soul, life, and heart and the woman its body. This old model of inheritance underlies the common — but by no means universal — convention that wife and children should share a man's surname. Such cultural atavisms notwithstanding, both parents contribute about the same amount of information to each newly conceived child; if anything, the mother's contribution is more important since she contributes a whole egg cell to the next generation while the father contributes only the nucleus of his sperm cell.

Immediately after an egg cell takes in the head of a sperm, chromosomes from the two parents combine to form a new version of the human genome, and the egg cell is transformed into the first cell of what will be a person. The new genome in a fertilized egg is not the new person nor even a coded version of the person; it is the archive of information that the developmental clock will use to form a new and genetically unique body from the descendants of one fertilized egg cell. The products of certain genes — called regulatory proteins — attach to the opening stretches of other genes in the genome, turning them on or shutting them down, giving the cell a new protein or taking one away. When the new protein is itself able to turn other genes on or off, it sets off a cascade of gene-switching. Eventually, gene circuits end up conferring a specialized, differentiated fate on every cell of the embryo. The fertilized egg begins this cascade of differential gene expression by opening a new human genome's genes in precisely the right way so that a baby emerges a few months later. We are thus all born of women in a second, deeper way: all of us get our start through the action of proteins present only inside our mother's egg cell.

The cells of the body that differentiate into either sperm or egg cells are called germ cells. Human germ cells thus produce the sole transmitters of a human genome. With the exception of any sperm or egg cells that have succeeded in fertilizing each other, a death means the end of all inner developmental times for a particular version of the human genome and for the clonal population of cells it constructed from a fertilized egg to be a person. One-celled microbes pass on an exact copy of their genetic information by copying their genomes and then splitting themselves in two. For them, developmental time is frozen, for their "body" is always complete. The lines of germ cells passed on from generation to generation by our species resemble these lines of single-celled life. The resemblance is deep and old: germ cells escape death because the developmental clock is frozen for them as well. At the instant of fertilization, when developmental time begins for the other, mortal, cells of the body, the genes of the germ line are left alone, undisturbed and unused, preserved instead for a proper cascade of differentiated readings in the next generation. In all other cells of the body, the developmental clock continues to open and close different genes in different cells throughout a person's lifetime. For instance, sex hormones secreted by a small number of cells bring on puberty.[1]

The world outside the body can also change a cell's fate, by triggering a change in its choice of open genes. The senses are constantly sending the brain new information about the changing world, and sensing a change in the world outside actually changes the brain. When the nerve cells in our sensory organs respond to signals from the outside world — light, sound, touch — they do so initially by communicating with electrical and chemical impulses and then — in a delayed response necessary both for memory and for the stable wiring of the brain's circuitry — by altering the way they read their own genomes. The sensory experiences that wire up the brain in the first place keep rewiring it all through life. Much of what we think of as most ineffable about our brains — our conscious sense of the world around us, for instance, or our memories — are at least in part the consequence of nerve cells continuing to change one another's patterns of gene expression.

Make two fists and bring them together to appreciate the rough volume and shape of the human brain. Its unprepossessing appearance has led to many deprecating descriptions; my favorite is the mathematician Roger Penrose's: a bowl of porridge. In its two wrinkled, wet, warm hemispheres lie chemical and electric circuits of the greatest known complexity and density in the universe. Aside from its shocking smallness — somehow I had always imagined the brain as far bigger than the skull that encloses it — the brain's other surprising aspect is its crumpled appearance. The brain is squeezed like badly packed clothes into its bony case. Any slice through it shows why: its outermost quarter inch or so is a continuous layer of densely packed and interconnected nerve cells so large that to fit into the skull it must be deeply folded and pleated, leaving about a third of its pinkish-gray surface showing when the skull is lifted off. This gray matter is the brain's cerebral cortex. Nerve cells in the cortex are organized in columns that run perpendicular to its wrinkled outer surface, making that surface a fine-grained intarsia or mosaic of the tops of many tiny cortical columns. Nerve cells from each sensory tissue send their information to different tiles of the cortex's mosaic. Adjacent columns then share the information with each other and with organized clusters of nerve cells elsewhere in the cortex as well as in the rest of the brain.

A wrinkled cortex is the product of two conflicting needs: first, the embryonic brain needs to develop for an extended time in the controlled and safe environment of the uterus, and second, after birth the brain has to respond in ever more subtle ways to the vicissitudes of a constantly changing world. The mammalian trick of nurturing an embryo in the mother's body limits a newborn's skull to the size of a stretched pelvic passageway, while the mammalian knack for complex behavior places a premium on a cortex of the largest possible area and therefore the greatest capacity to process information from its senses. The combination leads to the cortex that must be crumpled up to fit in the skull. Humans are one of the two mammalian species with the largest and therefore the most convoluted cortexes; the other — with a cortex even more complicated and folded than ours — is the highly social, handless, but not speechless porpoise.

Within one cubic millimeter — the size of a large grain of sand or a rather small diamond — the cerebral cortex contains about one hundred thousand nerve cells. Each nerve cell makes tens of thousands of connections to other nerve cells; the nerve cells in a sand grain of cortex make about a billion connections with one another and with more distant nerve cells as well. Connections from the cortex to distant parts of the brain and spinal cord are wrapped in a fatty sheath called myelin. Much of the inner part of the brain is called white matter because myelin has a milky appearance rather than the gray of the cortex's nerve cells. The other parts of the brain connected to the cortex through the white matter — called centers or nuclei — are also organized in aggregates of tightly wired clusters of nerve cells.

All signals between nerve cells, and all signals from the outside world, enter through a membrane. The membrane of a nerve cell is a fairly impermeable, fatty skin, penetrated by doughnuts and rivets of protein that stud it like the peepholes and locks in a New York apartment's front door. Nerve cells signal one another by releasing small molecules called neurotransmitters from their tips. When a neurotransmitter released by one nerve cell arrives at a nearby nerve cell's receptors and fits one of them properly, other portals in the recipient's membrane open and shut, turning on an electric current in the form of a flow of salts that travels the length of the nerve, triggering the subsequent release of neurotransmitters at its tips. The genes that encode the various gates, pores, channels, and pumps of the nerve cell membrane control which neurotransmitters will be able to trigger an impulse, which salts will flow inward and outward to propagate the electric pulse to the nerve cell's tips, and which other neurotransmitters will be sent out by those tips to other nerve cells in turn.

Embryonic cells become nerve cells by the activation of a set of genes whose products allow them to receive and transmit chemical and electrical impulses at their tips, just as other embryonic cells become pancreatic islets by the activation of another set of genes, including the gene for the hormone insulin. The amount of hormone produced by a pancreatic cell's insulin gene can be tuned by the responses of other cells to secreted insulin. In a similar way, the intensity and pattern of neural communication can affect the strength of the connections that nerve cells establish at their tips. A network of nerve cells is established when connections among a group of cells are tightened by gene activation. The almost simultaneous arrival of nerve impulses from many cells activates genes in the recipient nerve cell; they direct the production of proteins that then hard-wire the cell into a network with the cells that sent the signals. Connections are not hard-wired unless multiple impulses arrive within less than a thousandth of a second, as measured by one of the brain's internal clocks, the nerve cell's coincidence clock.[2]

The circuits in our brains are thus assembled through the coincidence clocks of nerve cells after the ubiquitous developmental clock of differential gene expression lays out a rough draft of the wiring circuitry in the embryo. In the early embryo's brain, there are a vast excess of weak connections among nerve cells. As rough, even random, clusters of nerve cells hook up to one another, the embryonic brain buzzes with cross talk until impulses begin to arrive from many different cells in a thousandth of a second or less. Once recipient cells harden their active connections with the cells that send these coincidental signals, any signal from even one cell in the network will get through. If multiple coincidental inputs do not harden the initial embryonic connection, it will dissolve. Only after a nerve cell gets wired into functional circuits can it properly respond to the hormones that keep it alive; if it does not end up with a large enough set of stable connections to other nerve cells, a brain cell will usually die.

As soon as the fetus begins to be exposed to the world through its senses, its sensory nerves begin to signal the brain, allowing the coincidence clocks of the brain's nerve cells to begin to sculpt the functional connections to each of the senses. The senses and the way the brain interprets their signals are thus both products of the cooperative interaction of the developmental and coincidence-counting clocks; each nerve cell's embryonic capacity to sense time's passage at each of its tips creates and gives meaning to the senses.

Coincidence continues to resculpt the brain's circuitry throughout our lives as experiences continue to add to the establishment of connections among nerve cells. The size and complexity of a child's brain increases from birth until about the tenth year as the coincidence clock continues to maintain an ever larger number of connections among an ever larger number of nerve cells. Thereafter, though, the clock serves more as a winnower than a seeder as nerve cells in the brain begin to die back for want of sufficient new connections. By late adolescence, connections in the brain and numbers of brain cells are reduced to the level of a two-year-old's; from then on they continue to decline slowly for the rest of life. A considerable portion of each brain's final circuitry is created by experiences rather than genes. Since every child's experiences will differ from every other's, the wiring patterns of any two brains — even those of identical twins — will be entirely different from infancy on.

The eyes may see and the nose may sniff the air, but the brain is in odorless darkness, its networks of nerve cells completely secluded inside the skull. Five centers deep in the brain unconsciously process sensory information so that it can become part of the conscious recognition of the world. At the very base of the brain, beneath the white matter and on top of the spinal cord, sits a knob of densely packed neurons called the thalamus, organized, like the cortex, in a sheet of closely packed columns. We know the thalamus is critical to consciousness, since even the smallest damage by stroke or injury will leave a person alive but in a profound coma. Both the thalamus and the cortex connect in turn to four other major centers that are involved in making sense of the world. Studies of patients who have suffered damage to these centers suggest that one — the hippocampus — is critical for the storage of memories and that another — the amygdala — is central to a person's emotional state.

The third center — the medulla — organizes movements like walking and breathing, which may reach consciousness at any given time; the fourth — the cerebellum — is a sort of second brain, like the navel of a navel orange. It sits at the top of the spinal cord, tucked behind and below the cortex, and connects directly to the spinal cord. The cerebellum takes signals from the cortex and sends its own signals through the spinal cord to the limbs to maintain steady, controlled movements. The cerebellum is essential to movements involving conscious discrimination: it is more active when you pick up the correct change, say, than when you grab a tossed ball. The cortex, thalamus, hippocampus, medulla, and cerebellum all share signals coming into the brain from the senses. They link these signals with memories and feelings and use the mix of past and present to guide the animation of the body. It takes some time for the brain to comprehend the information it receives from the senses. That time is taken up in converting the signals generated by the outside world into nerve cell responses that can spread throughout the brain.

The sense of smell is a good place to start an examination of the past's presence in our conscious grasp of the world. It is the oldest sense and the one that reaches most directly into the brain. Odors trigger an early warning system in the brain which keeps us away from such airborne dangers as fire — where there's smoke, there's fire — and attractive but bad-smelling food. By sniffing, our mammalian ancestors could make life-saving choices without exposing themselves to the risk of a nibble. Picking out chemicals from the stuff in our mouths as well as from the air we sniff, our sense of smell gives food any taste it has beyond saltiness, bitterness, sweetness, and sourness, the tongue's four ancient signals of blood, poison, calories, and unripeness.

When we sniff or chew, we first dissolve a mixture of airborne chemicals and bring the solution to a space behind our nose and above our palate, the retronasal passage. There the chemicals arrive at an odor-sensitive, dime-size dollhouse carpet called the olfactory epithelium, made of about ten million nerve cells. Each cell's membrane is covered with a protein that can recognize and bind to one or more dissolved chemicals, or odorants. Proteins just inside an odor-sensitive nerve cell's membrane respond to the binding of an odorant by sending an electrical signal along the length of the nerve cell to its tip. There, the cell releases neurotransmitters that jump to the receiving tips of nerve cells from the brain. These brain cells — themselves quite insensitive to odors — respond to electrical signals from the olfactory epithelium by firing their own electrical and chemical signals. To our conscious mind, also a product of the silent, odorant-free brain, this set of brain signals is our sense that we have smelled something.

Mammals, each with no more than about sixty thousand different genes in their chromosomes, give over at least a thousand genes — almost two percent — to the coding of different odor-receptor proteins. The sense of smell was far more important to the survival of our ancestral species than it is to us today. Our germ line's inability to give up such a profligate commitment of genetic information is an example of how natural selection differs from conscious design. There are many more perceived odors than even the thousands of different receptor molecules; more complex smells probably begin as mixtures of odorant molecules that bind to subsets of different receptors. Professional testers of perfume, coffee, and the like can distinguish among hundreds of thousands of different odors, and any of us can tell the difference among tens of thousands of different-smelling things. As the olfactory epithelium connects itself to the brain during early development, it uses the past in an unexpected way. While objective time may never reverse itself, a close look at the different ways these odor-receptor proteins are used reveals that the developmental clock of olfaction harkens back in adulthood to embryonic development, fertilization, and even before.

The olfactory epithelium senses the world of smell through a thousand different sets of about ten thousand receptor cells each. For this to work, each set of cells must first run back into a distinct region of the brain's olfactory processing center. But the olfactory epithelium is not a formal garden of clearly marked, separate patches of nerve cells, each patch bearing a different odor receptor. Rather, it is a meadow strewn with nerve cells carrying different receptors, mixed together at random. Thus there is no direct correlation between a region of the olfactory epithelium and a region of the brain. This raises an interesting problem: the genome has already spent a few percent of its genes on odor receptor proteins. It cannot afford to also encode separate pathways to the brain for ten million olfactory receptor nerve cells. The solution found by natural selection involves a distortion of the internal time kept by the developmental clock, so that each time we sniff or taste the world we are transmitted back to an earlier time, when the embryos of our bodies first formed.

An embryonic nerve cell born in the olfactory epithelium first makes a choice from its thousands of odor receptor genes and puts only the protein receptor coded by that gene into its membrane. As the growing olfactory nerve then extends itself toward the brain, its choice of odor receptor protein steers it to its right place. The receptor protein does this by recognizing and binding to a particular track of fixed, odorant-like molecules laid down earlier by other cells in the growing brain. Like ants following an odor trail on the floor of a forest, embryonic olfactory nerve cells covered with the same odor receptor follow the same path to the same place in the brain. In this way, odor receptor proteins actually allow the olfactory epithelium to be hard-wired to the brain so that later in life — after the baby's first breath — distinct odors will trigger specific parts of the brain.

Of the two times the odor receptor proteins function, the unconscious, embryonic one is clearly the more critical, since without it there would be no circuitry of smell at all. Thus, a single set of genes and their proteins first organize the brain for future olfaction in the embryo, then operate the sense of smell in the adult organism. But as a result of the first, unconscious, embryonic use of odorant receptors, the world we can smell is a very peculiar portion of the entire world of possible smells. If a chemical cannot interact with a receptor designed to build our sense of smell in utero, we cannot smell it later in life. It is quite strange to realize that our sense of smell is constrained by the embryonic function from ever completely attaining the adult one.

A small number of odor receptors have additional internal meanings, one of which goes back even earlier in embryonic time, to the frozen time of the germ line itself. Sitting in the membrane behind the sperm cell's DNA-packed head are a ring of odor receptor proteins. Recent work suggests that these particular odor receptor proteins are helmsmen of the sperm, converting a signal from the outside of the sperm cell — a molecule secreted by an egg cell — into a change in the direction of the sperm tail propeller. Apparently the same molecular hide-and-seek that will later lay down the neural pathways from the olfactory epithelium to the brain and, later still, present the mature brain with a universe of odors first operates to help the sperm get to the egg. When we smell some things, we are thus inadvertently remembering the construction of at least a part of our brain and, possibly, even our beginning.

It is plausible that some odor receptors play an even earlier role in the urge to mate, the behavior that leads to fertilization in the first place. Rodents have a second organ of smell in their heads, called the Vomero-nasal organ, or VNO. The olfactory nerve cells of the VNO run to a different, separate part of the brain, which connects directly to regions responsible for muscular activities of various sorts. Odorants that cause an instinctive and stereotypical response are called pheromones. It seems clear that rodent pheromones exert their actions through the VNO. The two major behavioral responses are a suckling response when the stimulant comes from milk and a mating response in the male when it comes from a receptive female.

We have a rudimentary VNO behind the wall that separates the two nostrils, but it is not clear whether it functions as a sensory organ or whether it is just an atavistic gift from some ancient common ancestor of rodents and primates. Since sexual arousal would be the most likely behavioral output if the VNO were stimulated, perfume makers have begun to search for chemicals that change the electrical properties of VNO nerve cells. The VNO is not wired to the part of the brain that processes the sense of smell, so although such artificial human pheromones might make a perfume work in extraordinary ways, they would not by themselves be consciously perceived as odors. If any human odor receptor proteins do function in the VNO as well as in the olfactory epithelium, then the smell of some odorant yet to be isolated would be the molecular recollection of intimate events leading to one's conception.

Assembled from a single cell by the developmental clock, all our senses depend on the past. As the senses get wired to the brain, nerve cells draw on an initial set of meanings for their receptor molecules, which precede any meanings we may give them by association with the sensations they may signal later in life. The way our eyes and brains perceive differences in color, for example, is as dependent on the past as our perception of smell. Our capacity to distinguish among the colors we call blue, yellow, green, and red may seem to bring us aspects of the outside world with total clarity and objectivity. Many of the differences in color we can see, however, are not the products of rational design but of ancient genetic changes fixed in the human germ line millions of years ago.

The first scientist to address the problem of color perception was Isaac Newton, whose seventeenth-century studies on "Opticks" remain an unmatched triumph of experimental simplicity and conceptual subtlety. Using only prisms and sunlight, Newton showed that white light was a mixture of separable colors and that the colors resided in our perception, not in the lights themselves. Newton saw the splay of hues made by his prisms as seven distinct colors — red, orange, yellow, green, blue, indigo, and violet — and he took great satisfaction in finding seven whole notes — a celestial octave — in the prismatic spectrum of the sun.

A century later Thomas Young — the same man who saw how to decipher the Rosetta Stone — first showed that light had the properties of a wave by capturing the wavelike interference fringes cast by light as it passed through two close, narrow slits. He used sunlight and prisms in a series of classic studies that led him to conclude, more or less correctly, that our eyes perceive color through three different kinds of nerves, each most sensitive to a different part of the visible solar spectrum. Later still, about a century ago, James Clerk Maxwell completed the experimental deconstruction of color. He was a great organizer of natural phenomena who also reduced the complexity of all electromagnetic radiation, from light to radio to X-rays, to a single set of beautiful equations. Passing a beam of sunlight through a box of prisms with movable slits, he was able to match any color of light by mixing spectral, monochromatic red, green, and blue lights. Yellow light was peculiar: it could be generated by a mixture of red and green light of equal intensity but also by the single spectral yellow lying between them.

Today we know that to distinguish colors, our brains have to make sense of a constantly changing rainstorm of weightless particles called photons as they fall on a few hundred million nerve cells at the back of our eyes. These nerve cells are part of a meshwork called the retina. The light-absorbing cells of the retina are called rods and cones; there are three kinds of cones, but only one kind of rod cell. The cells are named after the shapes of their stacks of receptors; each rod and cone make so many copies of its light-sensitive protein that the membrane holding them becomes folded and pleated with the load. The drops of light our rods and cones pick up can be very few and far between: at the limits of vision, we have a fifty-fifty chance of perceiving a photon drizzle whose total energy is less than one thousand-million-millionth of a watt, about as bright as the light shed by a candle seen from a distance of ten miles.

Though we are sensitive to hundreds of different levels of brightness above that limit, the range of visible energies, or wavelengths, of light falls inside about a factor of two from longest to shortest wavelength, the very octave Newton intuited. In comparison, we can hear a tonal range of at least seven octaves; the boundaries of the visible octave are set by the absorption of light by the atmosphere and by fluid in the eyeball. We pull a very large number of gradations of color out of this single octave of wavelengths. Each gradation can be distinguished from the others by its hue, saturation, and brightness. Hue is what we usually mean when we speak of color; most people can distinguish about two hundred hues. Saturation is the degree to which a color rises above or sinks below the overall level of brightness of the surrounding field. Most people can distinguish about twenty degrees of saturation for each hue. As our rods and cones can receive hundreds of levels of brightness for each combination of hue and saturation, most of us see a world of about two million gradations of color.[3]

•

The brain can no more see light than it can smell odors. Before anything we call vision can occur, light-sensitive cells in the retina must get their information to blind cells in the brain. Unlike smell, color perception cannot work through the direct recognition of different molecular shapes by a large family of membrane receptors because light has no molecular form. To give the brain the information it needs to resolve different colors, small groups of rods and cones deliver their information to nerve cells in the retina. The retinal cells then either add or subtract these signals from the clusters of light-sensitive rods and cones, then send the result of their calculation down the optic nerve to the part of the cortex at the back of the head, just beneath the skull.

At each interchange along the way, color-coding retinal cell impulses combine with other retinal signals, some representing the overall brightness of the image, others the edges of things seen, and still others their movements or relative distance from the eye. Together, the colors, boundaries, and movements of the scene merge in the brain to form the mind's smooth, conscious perception of a colorful, fluidly changing world. Signals from the rods and cones are involved in all aspects of vision; it is the retina that breaks the signals down into the information the brain needs to abstract the many different aspects of full vision.

The hundred million or so rod cells work best in the dimmest light, where cones fail. Rods are connected — in clusters of ten or so — to single retinal cells that pool and amplify their signals so that even when only one rod out of ten is stimulated, the brain will be notified that light has arrived. From this evidence, the brain will assemble a blurry, colorless image; that is why all cats are gray at night. The cone cells of a retina are fewer in number — about ten million in all — and most are packed closely together at the center of the retina. Hold up one finger, stretch out your arm, and look at the fingernail: it is being projected by your eyes onto the central patch in each retina that is the most densely packed with cone cells. This region sends the brain information of the finest resolution, both of detail and color. When we look closely at something — like the page you are reading or the screen I am looking at as I write — we are sweeping its image over these retinal patches. Cones become sparser farther from the patch until they are all but missing at the rim of the retina. That is why peripheral vision — the brain's processing of information gathered by photoreceptors at the outer edge of the retina — is very poor at color recognition, even in bright light. When you sense that a bird has flown by, too fast and too far off to the side to be seen in any detail, the cone-poor rim of each retina has actually informed your brain only of a colorless, moving object.

Olfactory nerves quickly tire of signaling the presence of a persistent odorant, so we get used to a bad smell even when the stinking material still lies before us. Our retinas also quickly tire if their visual fields are kept the same for long: when an image becomes stable on the retina, it disappears. To keep the world before us, an unconscious pendulum drives the muscles of the eye to constantly sweep back and forth. This scanning oscillation flicks our eyes at a constant rate so that regardless of the scene, each rod and each cone will send out a background signal timed to the sweep of the eyes. Though this background wave reaches the visual cortex, it does not reach consciousness. The brain's perception of the world — the smoothly changing world we see — is built up from perturbations of this background signal generated when the retina is exposed to variations in light and dark that change, not only because of the sweep of the eyes, but because of real changes in the view.

At the same time as it is analyzing the signals coming from different regions of the retina, the brain is also adding up the entire retinal signal to give it an idea of the intensity and color of the light that is illuminating the entire visual field. This information about the quality of the ambient light is then used to interpret the meaning of the particular set of color-specific signals sent from any region of the retina; as a result, an object's perceived color remains constant even if the overall hue of the illuminating light changes. This is why the page you are reading will look white at noon, in a sunset, under a light bulb, or by candlelight, even though the actual spectrum of light reflected from it will be different in each case.

For at least the last thirty million years, this constant global renormalization by the brain has allowed us and our ancestral primates to make sense of the colored world. Despite the skittering of the eyeball, the constant shifts from bright light to shadow in a dappled forest, and the change in hue of the sun from dawn through noon to dusk and sunset, a page like this one is always white, and the leaves of a tree are usually greener than their ripe fruit. The conservation of color — the final interpretive step between photon absorption and color vision — is essential to our sense of the physical continuity of objects over time. Without it we would, I think, go quite mad, or at least wish to keep our eyes firmly shut.

All three kinds of cones can absorb light from almost the entire visual octave, but the pigment in each absorbs a different fraction of the octave with greatest efficiency. No single cone by itself can be responsible for any perceived color. Any one type of cone cell would not be able to signal the difference between a bright light at a wavelength its pigment can absorb only poorly and a dimmer light at the wavelength its pigment absorbs best. The different light receptor proteins in the three sorts of cones are named for the wavelength of light each absorbs most efficiently. Cones that carry the short wavelength receptor protein absorb light best from the shorter half of the visual octave. Cones that carry the medium wavelength receptor or the long wavelength receptor both absorb light best in the longer half of the visible octave; the actual wavelengths they absorb best differ by only about a tenth of an octave.

To see a color, the brain must get the combined output of at least two different cone cells. Additive and subtractive nerve cells in the retina link cone cells with the visual region of the cortex of the brain. They give the brain the information it needs to perceive color by sending signals that represent either the sum or the difference of outputs from two adjacent cone cells. A subtractive retinal cell is silent when both its cone cells send out signals of equal strength; it signals the brain only when one of the pair is more strongly stimulated by incoming light than the other. A signal from a subtractive retinal cell cannot tell the visual cortex whether the scene is light or dark, only whether the incoming light at that point on the retina is equally efficient at stimulating a pair of cone cells. The additive retinal cell connected to the same pair of cones fills in the cortex with the missing brightness data by signaling in proportion to the total amount of light absorbed by both its cones.

Some of the colors we see from the longer half of the visual octave — the greens and reds — come from a particular circuitry involving two subtractive retinal cells and one additive one for each set of one medium wavelength cone and one long wavelength cone. The additive retinal cell informs the brain of the brightness of the light striking the pair of cones. One subtractive cell sends a signal to the brain when its pair of cones is exposed to wavelengths that trigger a stronger output from the long wavelength cone than from the medium wavelength cone — that is, longer wavelengths. The brain reads this subtractive signal as the color red. The other subtractive retinal cell signals the brain when light triggers a stronger output from the medium wavelength cone than from the long wavelength cone; that is the brain's cue for the color green.

The third type of cone absorbs the short wavelengths of the visible octave best; it is barely able to absorb light from the longer half of the visible spectrum. It too is wired to a set of two subtractive and one additive retinal cells. These retinal cells do not connect to another cone but to a nearby additive retinal cell of the first sort, whose output represents the combined output of a nearby pair of long and medium wavelength cones. One subtractive retinal cell signals the brain when the short wavelength cone is active and the additive cell from the other pair of cones is silent. The brain interprets this signal as the color blue. The other subtractive retinal cell signals the brain when the short wavelength cone is silent and the long plus medium wavelength additive cell is active: that is the signal the brain reads as yellow, and that is why we see yellow when green and red lights are mixed. The third retinal cell of the set — the additive cell connected to a short wavelength cone and another additive cell — sends a signal to the brain only when all three cones are triggered: that is the brain's signal for white, and that is why we see white when blue and yellow lights are mixed.

The set of colors created by a double system of subtractive signals — blue vs. yellow and red vs. green — was invented many times. The first time was millions of years ago by natural selection; subsequent inventions include color film and color television. The screens of video monitors are mosaics of tiny patches that can be made to glow red, green, or blue. At a distance our eyes cannot resolve the separate dots, so our retinas merge them into mixtures of the three wavelengths. When all three pigments glow with equal brightness, they stimulate all three kinds of cone cells to the same extent, and we see that as white, gray, or black.

Every TV set reproduces both the sweep of the eye's background oscillations and the brain's use of difference circuits to generate a colorful, changing scene. A TV has enough electronic brainpower to combine the transmitter's three abstract difference signals into an electronic wave. As it sweeps a beam of electrons across the inner surface of the video screen about sixty times a second, the beam makes the red, green, and blue dots of phosphorescent minerals glow or not, as required by the scene. TV screens need no yellow dots for the same reason our eyes have no fourth class of cones: a yellow image triggers the TV camera's first difference circuit and not the second. You see no yellow dots when the video is showing you a pat of butter; red and green dots glow and blue ones are dark, but you see yellow.[4]

Color vision is not a calibrated, smooth response to the entire visual octave nor even a fair sampling of all its possibilities. Retinal subtractions set up either-or choices in the perception of colors. It is not possible to imagine a bluish-yellow or a yellowish-blue because the wavelengths that stimulate all three cones equally are seen as white; neither a reddish-green nor a greenish-red can ever replace yellow in our brains. We never see colors of this kind because, in every case, seeing one color in the pair is precluded by seeing the other. This wholly arbitrary gap in our range of colors, and our consequent inability even to imagine what ought to be perfectly reasonable additional colors, are historically determined limitations to our senses, expressions in our bodies of events that took place in the ancient past.

The longer wavelengths of the visible spectrum are remarkably richer in perceived colors than the shorter ones for other historical reasons. The way the past determines this aspect of our sense of color was first understood less than a century ago by the American psychologist Christine Ladd-Franklin. Her work began with a comparison of visual circuitry in humans and other mammals. Most other mammals see colors, but in an even narrower range than we do. All mammals have a cone cell quite similar to our short wavelength cone, and all have the neural circuitry that distinguishes colors by subtracting signals from pairs of different cones. But most other mammals lack a third cone; their additive and subtractive retinal cells link short wavelength cone cells directly to a second cone cell absorbing longer-wavelength light. Their retinas can send only one set of subtractive signals to their visual cortex. This signal enables their brains to distinguish between lights of shorter and longer wavelengths, but it does not give them the capacity our third cone cell and second set of subtractive cells give us, to see many additional colors in a small part of the longer portion of the visual octave.

When a retinal cell in the eye of a typical mammal receives a weak signal from the short wavelength cone and a strong one from the longer-wavelength cone, the animal sees some color that is certainly different from blue. We may be tempted to call it yellow, but that is an anthropomorphism: it is no color at all that we know, but rather a "nonblue" encompassing all reds, yellows, and greens. It may be approximated by a range of tans and browns, but it is really beyond our regular color vocabulary. When a cat looks at the noonday sky, for example, pulses from subtractive cells in its retina tell its brain to see light of a cool, bluish color. If the cat then were to look at a patch of grass or a buttercup, or blood from a wound it may have just inflicted on a mouse, it would see the three scenes in shades of the same ineffable nonblue.

Humans and Old World monkeys are today's living descendants of the first mammal to develop a third functional cone cell. Our third cone, and the circuitry that connects pairs of long and medium wavelength cones through retinal cells, began as genetic changes in the germ cells of a member of a now extinct species of primate. While we do not know the details of these genetic changes, it is clear that a gene encoding the longer-wavelength light-sensitive protein was accidentally copied twice, leaving one copy next to the other in its X chromosome. In time, the two initially identical copies of the gene became different as each accumulated a different set of mutations. The descendants of this primate apparently benefited from some of the mutations, for within a relatively short period — a few million years — natural selection had chosen primates with particular, new versions of the duplicated genes. Cone cells expressing either one of these genes or the other became the ancestors of the medium and short wavelength cone cells in our retinas. Once these new pairs of cones captured the retinal color discrimination circuitry of the blue/not-blue system by becoming linked to each other through two opposing subtractive retinal cells and one additive one, our ancestral species was able to see the differences we call red, yellow, and green. Today as then, when both new cones are equally stimulated by incoming light, the ancient system takes over, and the brain interprets the retinal cell signals in the ancient way as nonblue; with our second, new system in place, we can see that color as yellow.

The colors we see, and the ones we don't see — like the hundreds of colors that we might have seen between blue and green if the new, third cone were more sensitive to the shorter-wavelength end of the visible octave — are ancient choices made for us by natural selection. Because the genes of the two new cone proteins appeared by duplication so recently, they do not differ much. The wavelengths most efficiently absorbed by the new pair of cones differ only by a few percent, and at the molecular level the two light-sensitive proteins differ by less than one percent in overall structure. Only a few of a dozen or so molecular differences — three hydroxyl groups, six atoms' worth of difference out of about sixty thousand atoms each — are responsible for the entire difference between them. Yet without the inherited accumulation of these small differences, we would all be as colorblind as a cat.

The similarity of the two new genes also accounts for their instability: in about one percent of female germ cells, one of the two copies is lost. When that happens, a man is born with only two kinds of cone cells. It is usually a man, since the genes are part of the X chromosome, and men inherit only a single copy of the X chromosome from their mothers, while women inherit two. Though this situation is called color blindness, it is not a blindness at all but rather an atavism, a throwback to an earlier time in the history of our species and its ancestors. Men with only two types of cone cells see the world the way most other mammals see it.[5]

Understanding how the brain pulls the colors it does out of the visible octave of wavelengths still does not explain the selective advantage of being able to see red, yellow, green, and dozens of colors in between where most mammals see only various intensities of the same yellow-brownish nonblue. Why did the gene duplication that occurred in an obscure primate some thirty million years ago persist in that species, and why was it retained thereafter by dozens of primate species that evolved from it? One likely answer is mundane: food. As bees are to flowers, primates are to fruits. For nine tenths of the time since primates first appeared, the ones that were our ancestors have lived in trees, and their diet consisted primarily of the fruit of those trees. Even today, arboreal monkeys get most of their nourishment from fruit, and they are picky. Certainly, for these monkeys, being able to tell ripe from unripe fruit is of some advantage in surviving and raising young. Probably the same was true of the immediate descendants of the first primates to carry a functional extra photoreceptor gene. In this light — so to speak — it is not surprising that one of the few jobs a colorblind man cannot do well is picking ripe apples.

All perceived differences among the reds, the greens, the yellows, and their intermediate hues exist because our ancestors were able to reuse an ancient bit of retinal circuitry — difference signals — to differentiate new colors from two light-sensitive proteins that differed by only a few atoms. The accident of a gene duplication in an ancestor whose food had to ripen to red and yellow before it could be eaten, and not any objective allocation of colors by wavelength across the octave of visible wavelengths, gives our brains today the information we perceive as dozens of hues in the red-yellow-green range. It is precisely the absurd smallness of the difference between the two new cone proteins that allows us to distinguish colors that differ from one another, in terms of wavelengths, by less than one part in ten thousand. All the distinctions we make to see plaids, Kandinskys, traffic lights, and flowers are stimulated by wavelengths of light that are so close to one another, it is as if symphonies of music were being performed and perceived in the range of tones falling between a B and its nearest B-flat.

To identify an odor is not only to consciously remember the previous times one has smelled the same mix of odorants but also to unconsciously recall a set of earlier times: the time when one's brain was assembled, the time when one began as a fertilized egg, and even possibly the time before that, when one's parents found themselves unconsciously attracted to each other in part through an imperceptible odorant. In the same way, to see the colors from red to green is to recover a story from the unconscious evolutionary history of our species, when natural selection built on our ancestors' new ability to see red or yellow fruit against green leaves where other animals saw only one brown against another. If our species' ancestors had had a different history, we might be able to perceive an equally large number of different colors in the shorter half of the spectrum but only a few yellow-oranges in the longer half.

So much for the notion that our senses measure natural phenomena objectively. Instruments built to display every wavelength as a different color are called — correctly — false-color imagers. All other instruments of scientific measurement are necessarily false in the same way: they do not merely refine our senses, they overcome them, and they overcome the body's history in the process. Science's need to overcome the ancient peculiarities of the senses raises the next question: if our brains, in the same way, are the products of ancient clocks, is consciousness also a product of the past? The answer is clearly yes; consciousness is as far from being a natural instrument of science as are any of our senses. Just as the past has established that there are odors we cannot smell and colors we cannot see, there may be ideas we cannot consciously articulate. This realization raises a new set of questions in turn: can science come to terms with these restrictions? If so, how does it do so, and at what cost?

2

Consciousness

Strictly speaking there is no need for the hypothesis that the psychical systems are actually in *spatial* order. It would be sufficient if a fixed order were established by the fact that in a given psychical process, the excitation passes through the systems in a particular *temporal* order.

— Sigmund Freud, *The Interpretation of Dreams*

. . . But bid life seize the present?
It lives less in the present
Than in future always,
and less in both together
Than in the past. The present
Is too much for the senses,
Too crowding, too confusing —
Too present to imagine.

— Robert Frost, from *Carpe Diem*

UNTIL RECENTLY, the biology of consciousness was an oxymoron, and philosophers had carte blanche in guessing at the nature of rational thought. That is no longer the case: a new generation of scientists has queried the human brain — and by extension their own brains — by direct experimentation. Though some students of the mind — psychiatrists and psychologists — and some students of the brain — neurologists and neurobiologists — still believe their separate fields run on parallel tracks, scientists working at the border of these previously separate worlds have generated a model of the mind in which all mental states are brain states. They have also found strong evidence for the persistence of the past in all of the brain's operations: rational thoughts — and by extension all scientific agendas, even their own — are rearrangements of neural networks, changes in wiring and gene expression driven by the brain's inner clocks.

The clocks required for consciousness resemble the ones that operate in the senses. They build and maintain the ever-changing circuitry of consciousness, making and breaking connections among sets of nerve cells in the brain. They are not like any clocks of our invention; one can, for instance, suspend a second of time, as we will see in a moment. Because these clocks are rooted in the ancient origins of our species, the integration of sensory information into consciousness works under the same constraint as the senses do when they draw their data from the outside world in the first place. Neither operation is the product of rational design. Conscious thought is as tied to the past as any of the senses.

Some decades ago, a patient of the physician Benjamin Libet's lay on an operating table, his arm naked, the back of his skull removed, the brain beneath it exposed. His brain was not damaged but more open to the world than any conscious brain had ever been. Its exposed surface included one region of the cortex that receives signals from the skin of the arm. The man was not asleep. Locally anesthetized against the pain of the surgery and conscious of his situation, he had chosen to allow a tiny glass tube — an electrode — to be placed on the cortex. The doctors knew that a pinch to the arm would send a signal from nerves in the skin to that very spot on his exposed brain; by sending a mild pulse of electricity through the electrode, they could mimic the skin's signal.

When they pinched the man's forearm, he said, "Ouch." He

was sure he had said "ouch" immediately on being pinched, but instruments showed that about half a second had elapsed from the instant his arm was pinched until the time he responded. It took the nerve only about a hundredth of a second to send an electrical signal from the arm to the cortex, not nearly enough time to account for the lapse in the patient's response. Then, without pinching his arm at all, the doctors began the flow of a gentle current through the glass electrode to the spot on the surface of the brain that receives signals from the arm. If the current was turned on for less than half a second and then shut off, the man reported no sensation at all. When the electrical stimulation to his brain was maintained for more than half a second, he did report a new sensation in his forearm, though it did not feel quite like a real pinch. It was more of a tingle, and it lasted as long as the electrical current to the brain continued. Pinches to the arm and tingles to the cortex were then mixed and matched, uncovering an unexpected difference between perceived and measured — inner and outer — time: for the man and, by extension, for all of us, the conscious present is demonstrably about a half-second in the past.

The half-second that the patient failed to notice between the instant of the pinch and the instant of his "ouch," and the half second of electrical signaling to his brain that he also did not notice, were both lost in time. Even more interesting, if his forearm was first pinched and then the electrode turned on at any time in the next half-second, its electrical stimulation completely and forever masked the sensation of the pinch from reaching consciousness. Even though the skin's signal of the pinch had reached the brain in the first hundredth of a second, the volunteer never perceived that his forearm had been pinched. Instead he eventually reported — after a full second of brain stimulation — the forearm sensation generated by the electrode.

Recall that when he was only pinched, he thought the pinch

came at the time he perceived it. But if the response had been instantaneous, direct electrical stimulation to the brain in the half-second after the pinch ought not to have been able to override the message from the arm. In bringing sensation to full consciousness, the man's brain must have somehow held time in suspension so that the instant of sensation and the instant of its conscious perception would appear to be simultaneous. When the electrical stimulation aborted that process, the time that had been set aside was nevertheless still lost. This loss of time was confirmed when the forearm and brain signals were reversed so that the pinch came half a second after the beginning of brain stimulation. Because his brain displaced the time of a stimulation backward so that perception would seem to follow immediately the man reported the pinch first, followed by the tingle of a brain stimulation, even though the brain stimulation began before the pinch.

The time used by the brain to bring sensations together and merge them with memories and feelings is time forever lost to consciousness. A reflex — a bodily response to a sensation that does not require conscious processing — confirms the loss of time in consciousness. Reflexes respond in far less than a second; at most, they take about a tenth of a second. When you accidentally touch a hot stove or bump your elbow, nerves just under the skin — wired directly to the spinal cord — stimulate other nerves there to signal muscles to contract. In a tenth of a second or less, a reflex arc, as these nonconscious responses to stimuli are called, will generate a movement that will remove the hurting body part from the source of its pain. Only after a gap of about a second does the sensation get melded into the brain's other activities so that we notice we've moved — and why — only after the movement took place.

When is "now" for Libet's patient — or for any of us? None of us senses a half-second's passing between the instant someone pinches our forearm and the instant we sense the pinch. Tickle your own forearm. No second goes by before you feel it. Yet all human brains must take a half-second to integrate sensations into a conscious moment. This tells us that we are all controlled by an internal clock of consciousness that is set a half-second early. It gives us the sense of simultaneity, although anything we think we are experiencing now has actually happened a half-second or so earlier.

Our sense of the present instant, the sense of being awake and aware, requires the brain to connect the present with the past in a second, deeper way. In the missing half-second, the brain blends information received by the cortex from all its sensory organs with its own version of the past, its stored memories of earlier experiences, including its memories of earlier emotional states. Only then does consciousness emerge, with its smoothly changing perceptions of both the outside world and our inner frame of mind. The brain has many networks of nerve cells that work together to maintain one or another aspect of consciousness. They are linked to one another by chemical and electrical connections; for consciousness to occur, these networks must also be linked through time.

The conductor in charge of bringing the symphony of consciousness out of the brain's separate centers is a synchronizing wave of electrical activity that sweeps regularly through the brain, from behind the forehead to behind the nape, forty times each second. Though all portions of the brain are swept by this forty-cycle-per-second wave, we cannot hear its hum nor see its flicker, because conscious attention to even the shortest instant requires the integration of many cycles of network activity. This wave links the centers responsible for processing sensory information to one another as well as to other centers responsible for unconscious and conscious activities of the mind, in particular the amygdala, the hippocampus, and the frontal cortex, where, broadly speaking, emotional states are generated, long-term memories stored, and the intentions to speak and to act generated.

According to a new model that builds on these data, our senses, words, behaviors, unconscious mental processes, and subjective conscious thoughts are a set of changes in neural networks. These networks were formed in many stages: first during embryonic development, by the activation of a set of genes whose products helped assemble the embryonic brain; then by another set of brain-specific genes, activated by synchronization through the coincidence clock; then finally by the forty-cycle-per-second synchronizing wave. Until recently, the lack of tools to see inside the thinking brain left open an alternative, behaviorist model: that conscious thought is too subtle and evanescent to be explained by any actions of mere tissue, and so it must be an imaginary, not a real, phenomenon. So long as experimental evidence could not connect changes in the brain to either unconscious or conscious events unless those events were accompanied by behaviors, behaviorists argued that any hypothetical parts of the mind that did not manifest themselves in behaviors did not exist at all.

To study the brain with scientific rigor, behaviorists logically restricted their experiments to ones in which the brain was the source of measurable effects, that is, behaviors. Behaviorists thought — or, as they might have said, they behaved as if — choice, consciousness, imagination, and even thought itself were immaterial because they could not be measured as physical events within the brain. This may seem outlandish: surely in the very act of imagining this vision of the brain, they were proving it wrong? But the rules of science are strict. Mental states based on past experience — even ones that serve as guides to measurable behavior — may seem to be real, but to a scientist, something real must be reproducibly available for experimental measurement. If thoughts could not be detected by instruments, then to a strict constructionist of science they did not exist.[1]

Scientists who considered consciousness to be a brain state expected mental activities to be detected as activities of the brain once proper tools were available. But building the tools that would visualize an aspect of consciousness was not easy. Unless a thought resided in a small number of nerve cells, experiments on small groups of cells or even isolated parts of the brain would be unlikely to recover scientifically acceptable evidence of a thought as a change in the brain's overall circuitry. Coupled with the ethical barrier to invasive studies of the human brain — reproducing Libet's work with a healthy volunteer would be hard to justify today — this meant that any widely distributed brain activity associated with a mental state or a thought would have to be detected by instruments that could simultaneously monitor rapid changes in electrical activity at many places in the entire brain, all from outside the skull. That was not possible until very recently, so even as late as the early 1980s, many students of the brain still agreed with the psychologist Steven R. Harnad, who remained convinced that "only after the brain has determined what we will do does an illusion of conscious awareness arise, along with the mistaken belief we have made a choice or had control over our behavior." Then, in an ironic salute to the model of the mindless brain, the issue was resolved in the mid-1980s not by pure thought nor great insight but by a truly mindless tool, the computer.[2]

The computer allowed noninvasive tools to display very rapid changes in activity throughout the brain. These new tools visualized circuits fixed not only in space, but also in time. Consciousness and introspection could be seen and understood in terms of brain function as evanescent networks of brain cell clusters linked by the coincidence of their firing. The thousandth-of-a-second coincidence clock that helps the senses wire themselves to the embryonic brain is also used to set up the time-sensitive links between networks. The capacity of coincidence to fashion networks allows each nerve cell in the brain enough time to participate in many different networks so long as each network synchronously fires its round of impulses at a different instant. In circuits linked through the coincidence of their signals, a nerve cell anywhere throughout the entire brain may be functionally connected to thousands of other nerve cells in many different networks, each operating synchronously to within a few thousandths of a second, and each contributing its output to the others.

This evolved strategy — which vastly multiplies the functional capacity of nerve cell networks within the brain — requires a universal beat so that each cell may know at any particular instant which networks its impulses are serving. The forty-cycle-per-second clock described earlier serves this purpose: visualizations of the brain's global patterns of electrical activity display a forty-cycle-per-second electrical hum, the brain's forty-cycle-per-second clock. Apparently, the same conductor's baton beat that establishes functional links between sensory nerve cells and their different home bases on the cortex establishes and maintains functional connections between each region of the cortex and the various brain centers beneath it. A brain cell can be a member of many different neural networks, each sending its signals back and forth in one specific fraction of a fortieth of a second after the most recent passage of the forty-cycle-per-second pulse. The many networks distributed throughout the brain can each communicate this way forty times a second, firing at precisely the same amount of time after the global sweep passes by. Coupled with the capacity of nerve cells to tell time to the nearest thousandth of a second, the forty-cycle-per-second hum unites the entire brain, allowing its sensory inputs, cortical centers of abstract processing, inner centers of emotional affect and memory, and its signals to the body's muscles to function together as a single organ capable of conscious thought.

Just as the sense of smell depends on a set of odor receptor proteins whose first function is to allow nerves from the olfactory epithelium to reach their proper places in the brain, so day-to-day consciousness depends on a coincidence clock that the embryonic and newborn brain uses to wire up its first functional circuits. Well before consciousness, and for all conscious time thereafter, the coincidental arrival of impulses strengthens the connections between nerve cells, and the failure of coincidence allows connections to dissolve. The entire brain uses the forty-cycle beat and the capacity of coincidence to rewire networks and to constantly rebuild itself in response to news from outside the body.

The plasticity of connections between nerve cells can even lead to a large-scale remodeling of the sensory cortex. The whiskers of a rodent provide it with a sense of place we humans can get only from a mixture of sight and touch; each whisker of a mouse sends its information — how strongly it has been bent by contact — to a vertical column of synchronously firing neurons in the cortex called a barrel. Normally, the five rows of whiskers map neatly to five rows of barrels whose synchronous signaling can be seen on the surface of the exposed brain. Touching one whisker generates electrical activity in the cells of precisely one barrel. When a mouse is deprived of all but one whisker, touching that whisker stimulates signals not just in its barrel, but in regions around that barrel as well, regions normally triggered only by neighboring whisker hairs. In the absence of sensory input from other whiskers, signals from the one whisker that still functions are able to entrain additional parts of the cortex.

One of the least attractive metaphors for the brain is that it is the hardware of the mind's computer. This comparison is usually made with the clear implication that as time goes by, computers will meet and overtake our brains, for they will expand in complexity and in their capacity to handle information while our brains remain stuck inside our skulls. It is a metaphor that severely underestimates the brain's plasticity. Though the nerve cells in our brains do not grow much after we are born, their connections constantly form and reform. There is no brain hardware in the sense of permanent circuitry; the brain's "wiring" keeps changing in response to the lives we lead. The brains of identical twins look even more similar than the brains of two unrelated people, but the connections among the nerve cells in each twin's cortex are far more different than the ridges on their fingertips. The connecting circuits change every instant in both twins, just as they change in each of us.

When two people — even twins — think the same thought, different sets of synapses are likely to mark that thought inside each skull. Indeed, the only way one can imagine identical circuits in identical twins is for the twins to have had identical histories and thereby to have stored up identical memories. Because no two brains are the same and no brain ever has the same circuit twice, it is unlikely that consciousness will ever be successfully modeled by a hardware-driven technology, no matter how small, fast, or complicated.

The forty-cycle-per-second background hum was first detected with the standard instrument of neurology, the electroencephalograph, or EEG. But its role in binding together sensory information and memories was not understood until a few years ago, when a very sensitive variant of the instrument, called a magnetoencephalograph, or MEG, was developed. Unlike the EEG, an MEG can locate electrical changes to the nearest cubic millimeter of cerebral cortex and can time these changes to the nearest thousandth of a second. Early EEG studies had detected a forty-cycle-per-second component to the electrical activity of the brain, but they could resolve neither its source nor its relative phases in different portions of the brain. With the MEG, the forty-cycle-per-second brain wave was pinned down in time and space. It comes from two distinct clusters of nerve cells in the thalamus, deep in the brain.

Each cluster is an autonomous oscillator, sending a forty-cycle-per-second wave out along its extended fibers to all parts of the brain. Though the two thalamic clusters put out the same frequency of background hum, each serves a different function. The output from one allows the brain to bind together the body's ever-changing sensory inputs; the other's synchronizes the brain's internal workings. When the two thalamic oscillators work in synchrony, they bind the activities of nerve cell networks together in the centers of the brain responsible for sensation with those responsible for abstract thought, feeling, action, and memory, and consciousness emerges. Together, the coincidence clock and the two forty-cycle-per-second beats appear to moot the philosophers' mind-body problem, enabling the mind to emerge as an expression — a differentiated function — of the cells of the brain and nervous system. [3]

The first thalamic cluster sweeps a forty-cycle-per-second wave of electrical activity through the cerebral cortex. The wave begins at the front of the cortex, over our eyes and behind our forehead. Its peak then moves smoothly and swiftly beneath the top of our head, where the cortex is receiving information from our skin and muscles, to the back and side regions, where the cortex takes in signals from our eyes and ears. This first thalamic wave starts its sweep again every half-cycle, or eighty times each second. Because the whole brain contributes to consciousness, successive intervals shorter than about an eightieth of a second cannot be consciously perceived.

Because nerve cells in the brain communicate with one another — or not — in very short bursts that occur every fortieth of a second, these bursts can be synchronized by the thalamic signal generator's global sweep of the cortex. The sweep of the forty-cycle wave through the brain is like the sweep of sunrise or sunset over the surface of the earth. As the boundary between light and dark sweeps over the turning face of the planet, a longitudinal slice of people go to bed or wake up in phase with one another.

People living at different longitudes who go to bed at sunset will be responding to the same signal at different times; their bedtimes will be out of phase with one another. What is true for people on the planet is true for the different centers of nerve cells in each of our heads. Coincidence in the instant of arrival of a global signal, and not anatomical nor geographical nearness, synchronizes the behavior of people as it does the functional circuitry of the nerve cells in their heads. Just as no interval shorter than a whole day will sweep the entire planet of people through bedtime, conscious time cannot contain any instance shorter than an eightieth of a second, the time it takes for the entire cortex to be swept from front to back by a peak of the global, synchronizing, forty-cycle-per-second wave.

Shorter intervals cannot engage the entire brain because the thalamic center's global wave cannot intersect all the different regions of the cortex in less time than it takes to sweep through the entire brain from front to back. The brain's inability to separate images that arrive too close to one another allows us to see a series of still images as a moving picture. A coil of film stutters its way through the projector showing one still frame after another; as long as each frame is shown for about a twelfth of a second or less, we see the succession of pictures as one moving scene. What holds for vision also holds for perception through the other senses; when two equally short spaced strokes tickle us, we feel them as one.

Synchrony between the forty-cycle-per-second clocks of the brain and a sensory organ is essential for the stability of the wiring between the senses and the brain. If the synchrony is disrupted early in life, connections between that sense and the brain may be lost. People whose muscles are too weak to bring both eyes to bear on the same object may not be able to project comparable images on the retinas of their eyes. In this case, the brain receives a doubled, offset pair of visual fields; when these images are processed separately by the brain, one sees two of everything, a condition called diplopia. Newborn kittens can be intentionally given permanent diplopia by fitting them with eyeglasses containing prisms that prevent the two eyes from sending overlapping fields to the brain. At first, both retinas send information to the visual region of the cortex in proper packages of forty-cycle-per-second pulses, and both register with the cat's thalamic clock, resulting in a wholly confusing, doubled view of the world.

After a while the cat's brain dismisses one of the two retinal images. In order for the cat to have any sort of useful vision, it allows only one of the two retinal signals to be synchronized to the cortex. Signals from the other retina, uncoupled from the first thalamic clock, no longer reach consciousness to confuse the kitten. It's all well and good in the short run, but soon the information from the decoupled retina suffers the fate of all information sent through permanently unsynchronized nerves: its connections to the cortex die back. Thereafter, even if the glasses are removed, the cat cannot recover vision in one of its eyes; it has permanently lost the ability to see three-dimensionally.

People can have this condition as well: when for any reason the input from one retina does not become synchronized to the cortex early in life, the eye wanders in its socket thereafter, contributing nothing to the visual field. Wandering eye — amblyopia — confirms the critical importance of the thalamic-cortical forty-cycle-per-second sweep in integrating sensory input into consciousness. The signal from the wandering eye is full of information, but because it is not synchronized to the signal from the useful eye, the information it sends to the brain cannot be brought to consciousness. I have had diplopia for decades. Because each retina had been wired properly to my brain through reinforced, coincidental signaling before my diplopia began, it can be corrected with prisms that cancel the offset and bring the two visual fields into register with one another. It is still odd when I take off my glasses and see two of everything; I never fail to wonder which — or what — is real.

Even when sensations are bound together by the sweep of the first thalamic oscillator through the cortex, they need not reach consciousness. In fact, that is the fate of most of the information that reaches the brain. A sensation of a particular sort — a sight, a sound — enters consciousness and is noticed only when a sensory organ's signals are brought into synchrony with the second thalamic clock, the one that synchronizes the centers of the brain that need no sensory input, the ones responsible for emotional state and memory. So, for example, when we hear a click, a very brief input to our brains from the nerve cells in an ear interrupts and resets the brain's forty-cycle-per-second spontaneous sweep. For the following two and a half cycles, or about a fifteenth of a second, the entire brain's forty-cycle-per-second output is synchronized to the input from the auditory nerves; in that period, we focus on what we hear, rather than on what we are simultaneously seeing or smelling, because the auditory inputs synchronously interact with networks throughout the brain, binding the inputs of our ears to the rest of the brain, including the parts of the cortex that integrate, abstract, and name things.

When two short clicks are presented in less than a hundredth of a second, we do not hear them as separate clicks because the second click does not have time to reset the forty-cycle-per-second wave again; it is too close to the first. Instead, we hear a slightly different tone, because inputs of both clicks are bundled into one single unit of perception, albeit one that is different from the one generated by a single click. Spoken languages use differences in sound — phonemes — that last at least a hundredth of a second. We can distinguish different sounds in any language because even the shortest differences — like the difference between the explosive beginnings of a "p" or a "b" — take at least that amount of time. This threshold is no doubt set in all languages by the eightieth-of-a-second threshold for the binding of sensations. In a familiar language, sounds are converted to meaningful words and sentences without conscious effort. Until a language is learned well, a reader is obliged to bring each sound or letter to full consciousness; this uses many more cycles of cortical clock, slowing the understanding of what one is reading or hearing.[4]

The binding of networks by the two forty-cycle thalamic oscillators brings the more distant past of memory into every perception. The personal past also enters consciousness in the missing half-second described earlier. A small fraction of that time is sufficient for the first thalamic clock to synchronize to an external signal — a pinch on the forearm, for example — and for binding the information to other parts of the cortex. The additional time is necessary to establish a second synchronization and binding, in which the sensory signal and the first thalamic sweep are both synchronized to the second thalamic clock. Each perception we notice emerges only after this second synchronization, which connects the sensory system and the cortex to the parts of the brain that carry past memories and current feelings.

Because the full processing of a sensation involves binding the sensory input to the cortical oscillations that represent memory, the brain never responds in precisely the same way to a stimulus, even when that stimulus is precisely the same. The networks of nerve cells set up by the two thalamic oscillators are constantly changing in response to states of mind. Experiments have shown, for example, that signals from a monkey's retinas cannot alone establish synchronized connections between the visual cortex and other regions of the brain. Cells in the visual cortex will respond differently to identical retinal information depending on whether the monkey is paying attention to the visual image. Attention to a spot on a screen rather than the background can, for example, reverse the output of a nerve cell in the visual area of the cortex. As one scientist involved in this work put it, "The cortex creates an edited representation of the visual world that is dynamically modified to suit the immediate goals of the observer."

The phase-locked integration of networks in the brain through the synchronization of the two thalamic clocks can normally proceed quite well even in the absence of any new sensory input. On the other hand, if either of the two thalamic clocks is damaged — by stroke, injury, or surgery — a person loses consciousness and falls into a profound coma. When we are awake, the two thalamic clocks link the entire brain. At other times, when outside sensations are not being brought to consciousness — in a daydream, a dream, or a fugue state — the second thalamic clock's forty-cycle-per-second sweep continues to pulse through the brain, building and associating memories with one another, unperturbed by sensation and unmodified by the senses. This is why dreams can seem almost real: both consciousness and dreaming use the same thalamic clocks and many of the same neural networks. Consciousness integrates these networks with new sensory inputs, while dreaming uses sensory memories.

A sleeping, dreaming person's brain is still processing, binding, and interpreting its own stored information, so the minimal unit of time for the dreaming brain remains the same as it is in the wakened brain, at about one hundredth of a second. But because the brain's work while dreaming cannot be updated by sensory inputs, dreams are free from the constraints of external time. The same short click or pinch that resets the thalamic sweep so that we notice it while we are awake has no effect on the brain of a dreamer. The dreaming brain maintains its phase-setting, front-to-back sweeping of the brain's cortex and its linkage of that sweep to the phased outbursts of both the hippocampus and the amygdala, but it does not respond to an auditory signal by resetting the phase of the sweep.

Since thalamic oscillations are self-generated and do not depend on sensory inputs, wakefulness and dreaming sleep are quite similar. The similarity was confirmed when MEG analysis revealed that the brain is swept by the same forty-cycle-per-second waves during dreams, but not during other phases of sleep. In a dream, the thalamic forty-cycle-per-second synchronizer sweeps through the brain, binding the outputs of various portions of the cortex and linking them to the outputs of the hippocampal memory stores and the amygdala's emotional states, just as it does when we are awake. Sensory memories enter dreams when we are asleep the way new sensory inputs reach consciousness when we are awake: by resonance with the thalamic clocks. Awake or asleep, the brain is, as Rodolfo Llinas of New York University puts it, "a closed system emulating reality as delineated by the senses."

Anesthesia is an artificially induced state resembling sleep, but it differs from sleep in certain important ways. Most forms of anesthesia reduce the brain to dreamless sleep, disorganizing the forty-cycle-per-second sweep altogether. Other anesthesias are compatible with consciousness; anesthesiologists may use drugs that block pain and paralyze muscles, for example, without interrupting the sensory or thalamic oscillators. Treated with such a drug, a patient may appear to be completely out of this world while she is actually fully conscious of what is going on around her, hearing and feeling a procedure being carried out on her motionless body. Certain operations require both kinds of anesthesia; in these cases the only reliable way to determine whether a patient has inadvertently been reduced to a horrible state of paralytic but full consciousness is to administer a click and see whether her brain responds by resetting the phase of its internal forty-cycle-per-second oscillation. If the click evokes a change in the brain's oscillator, she is conscious; if not, she is properly anesthetized.

The possibility of being conscious but deprived of sensation brings us back to Libet's patient with his forearm pinched and his brain exposed. The past is always with him, and not just because conscious perception melds an immediate sensation with the brain's memories. Conscious perception, the very definition of "now" to his conscious mind, takes many dozen sweeps of the cortical-thalamic synchronizer — the missing half-second — in order to bring together inputs from the stimulus of his forearm with his brain's internal moods and stored memories. His sense of the present instant takes a half-second to be established, but that half-second is necessarily free of sensory input itself. It is like a dream, and as in a dream it has no conscious duration. That is why his conscious sense of "now" will always be a half-second later than the events that define it. In contrast, the electric current applied through a glass electrode to his exposed cortex had no forty-cycle-per-second modulation, so it did not resynchronize the thalamic timekeeper. Without resynchronization, the internal workings of the brain were unable to refer time back by half a second.

As time passes, each of our brains is constantly strengthening some connections and weakening others according to our experiences. Natural selection conferred on us the capacity to interact with one another and the world through a developmental clock that produces a brain whose neural networks are locked to one another through two internal timekeepers. The constantly changing, backward-looking internal representation of reality that we call consciousness is a byproduct of that set of evolutionary events. Perceived colors or odors — or scientific data — are not different from the thoughts, memories, and dreams they bring about; all are inventions of the brain, aspects of its obligation to use the past in order to interact with the external world. Because consciousness is a full linkage of the brain's parts through the linkage of the two thalamic sweeps with the various forty-cycle-per-second outputs of the senses, its version of reality, filtered first through natural selection and then through our own different, individual experiences, is the only reality any of us can know. As one neuroscientist puts it, consciousness is a fundamental quantity, not definable in terms of other subunits; it is like time, mass, or electric charge, not simply their product.

The silent half-second the brain uses to mix new and stored inputs together with its own internal neural cross-talk also allows the brain to build the introspective models of past and future. The recent success of science in explaining how the brain binds multiple perceptions and memories into consciousness through the interaction of the three forty-cycle timekeepers raises the question of how science itself works as an expression of brain function. Put in scientific terms, the question is: to what extent does scientific introspection depend on the unconscious memories of the scientist?

Imagining — making the models of the past and future with neither the benefit nor the burden of new sensory inputs — is just like dreaming. Stored past thoughts and experiences — memories — reappear in conscious introspection, as they do in dreams. In each case, memory emerges as the brain reconfigures itself to meld neural representations of past events with neural representations of imagined events that may or may not reflect actual events at all. The outside "real" time essential for science's objective measurements cannot even enter the mix of past and present that we experience as consciousness except as a consequence of the cortex's binding up a set of constantly changing sensory inputs.

The second thalamic oscillator deep in the brain has neither need nor use for objective time. This is why the brain can have a creative instant in which time seems to dilate to contain an entire symphony, or a brand-new idea, in what is only an instant of conscious, objective time. It is also why time's passage has no fixed place in dreams. There, too, even as the inward processes of the brain are linked by the second thalamic oscillator in the absence of changing sensory inputs, the passage of time cannot be registered.

In his 1947 book *What Is Life?* Erwin Schrödinger famously predicted the structure of DNA before its chemical composition was well understood by seeing that the material carrying genetic information would have to have apparently contradictory properties: crystalline stability for the sake of stable inheritance, but also the capacity to change and exist stably in many slightly different forms for the sake of genetic variation. The genetic material would have to be an aperiodic crystal, and DNA is precisely an aperiodic crystal, with a stable scaffolding of two backbones made of alternating units of sugar and phosphate wrapped around a wholly informational, aperiodic sequence of base pairs. Synchronized networks of nerve cells in the brain would have pleased Schrödinger; they are aperiodic crystals of time.

The forty-cycle-per-second background rhythm of the sensory and thalamic oscillators is the stable, periodic part of consciousness, much as the two phosphate-sugar backbones of any stretch of DNA give every DNA molecule the same double-helical shape, allowing the information it contains to be expressed and propagated indefinitely. The neuronal circuits synchronized to a particular phase in each wave of the forty-cycle-per-second background sweep are the aperiodic, informational part of consciousness. They change through time the way the sequence of base pairs changes throughout the length of a double-helical DNA encoding a gene. Thalamic oscillations carry the information of huge numbers of interdigitated neural networks through time embedded in their forty-cycle-per-second hum, just as the information packed in a DNA molecule as a set of specific sequences of base pairs is carried indefinitely through time by the same twisting, repeating double backbone.

The information of neural networks in the skull is aperiodic in time but stable in space, while the information of genes in every nucleus is aperiodic in space but stable through time. DNA, encoding the structures laid in place by the ancient clock of natural selection and assembled by the clocks of embryonic development and nerve cell coincidence detection, establishes the thalamic rhythms and so makes consciousness possible. DNA is the text of life, but the neural circuits linked by coincidence and brought to consciousness through the rhythm of the thalamic oscillators are its music.

As I sat in my study in Vermont writing this book, my wall clock would sound its bell from the next room every thirty minutes, alternately ringing in the hour, a number from one to twelve, or marking the half-hour with a single bell. Each single bell so caught my attention that I would often have to get up to see which half-hour it signaled before I could return to my writing. The clock's distracting ambiguity was doubly bothersome just after noon, when each single bell might be signaling any time between twelve-thirty and one-thirty. When the clock did not distract me, I worked without any consciousness of the passage of time. The writing itself, the focus on a choice of word, the daydreams, the fantasies, the creative moments of insight, the memories that came unbidden — all took place in a complex, partially unconscious mental world in many overlapping different times at once. The ringing of the ambiguous clock was so distracting because it collapsed all these internal mental times into the one time science recognizes, the quantifiable present instant of sequential time.

Strong emotions and focused attention coexist in our minds, though we often keep them as far apart as we can. Scientists are like everyone else in this regard, only more so. Science engages the whole mind of each scientist, both the conscious and the unconscious parts; fears, fantasies, dreams, and memories are as important to a scientist as any measurements or model. The famous intensity and passion with which scientists push away the tug of emotion in their work is an inadvertent but potent example of the richly emotional context in which science actually takes place. Until recently, the emotional content of science could be ignored by scientists and nonscientists alike. Now, however, scientists engaged in understanding how the human mind works have brought us to a moment of ironic discovery, when their own work has made it impossible to deny the presence of the unconscious in science.

The same science that invented tools that allow us to study the brain with focused, conscious rationality has done what neither any religion, nor psychoanalysis, nor philosophy had done before. It has found components of the human brain that create the emotional, memory-laden, unconscious components of the human mind and made them material for further scientific study. This is ironic because, as we have seen, the tools of science are poorly designed for studying the unconscious: the methods and instruments of science work only in objective time. Human minds, made up as they are of both conscious and unconscious parts, work in ways that bear little if any relation to the passage of external time and must lose track of what time it is — must not hear the bell of the clock — in order to do their work.

Every scientist has a sense of her intrinsic capacity to be completely objective as she creates models of nature based on the inputs of her senses, however greatly those senses may be enhanced by instrumentation. But that sense of objectivity is no more solidly based on the underlying reality of her brain's workings than her sense of color is based on the fair representation of all visible wavelengths. In both cases, the past pervades the present modeling of reality, attenuating objectivity. Three questions are thus raised by these new discoveries of science: Can science tell whether there is necessarily a hidden, unconscious component to its own operations? Would the discovery make a difference in the doing of science? And, if it did, would a better, more self-aware version of science emerge from acknowledging that discovery?

3

Memory and the Unconscious

. . . it still strikes me myself as strange that the case histories I should write should read like short stories and that, as one might say, they lack the serious stamp of science. I must console myself with the reflection that the nature of the subject is evidently responsible for this, rather than any preference of my own.

— Sigmund Freud, *Studies on Hysteria*

Who controls the past, controls the future. Who controls the present, controls the past.

— George Orwell, *1984*

WITHOUT THE SELECTIVE RECALL of past events, the current moment is incomprehensible. Memory — the retention of the past so that selected vignettes may later be converted into aspects of current, conscious perception — precedes both language and self-consciousness, directly aiding an organism's capacity to survive in a changing world: folding the past in with the perceptions of the present allows a creature to detect and focus its attention on what is new in its world. Memory is so bound up in consciousness that we tend to overlook its ancient origins and its ubiquitous presence in the animal world. We have already seen that our perceptions do not take place in the instant they seem to represent and that, instead, consciousness emerges from the immediate past in the fraction of the second it needs to integrate its perceptions. In that second the brain also integrates the "now" of external, sequential time with the more distant past of memory, making the entire lifetime of a person a source of yet another, different version of internal time.

Some stored memories may be only seconds old, like a phone number held in memory long enough to make the call. Other memories, ones laid down in our first months and years as infants, will be of emotions felt before we had words to describe them to ourselves. Among the centers of the brain that engage in unconscious mental activity are those that maintain stable but untapped neural networks representing the memories of all perceptions that have neither faded away nor pushed themselves into our waking minds. As we focus our conscious attention, we unconsciously sift our store of memories and bring to consciousness some, but not all, of the memories associated with that perception. Memories may also return to consciousness from the past at any time, without any new sensory input. A memory may sometimes be recalled in this way because it overlaps with an actual event, but a memory may also reach consciousness because the feelings evoked by a new event are similar to those associated in that memory with a different, earlier event.

Our emotional responses to an experience represent an unusual aspect of perception: they are as real a part of experience as any perception of sound, color, or smell, but they do not enter the brain through a single sensory organ as these do. An emotional perception emerges into consciousness as a feeling; the emotional content — the feelings that result from an experience — is called its affect. Just as the brain can store a memory of the colors or sounds associated with a past experience, so too can it store and recall an affect. We all instantly and permanently tie the erotic, frightening, satiating, fight-inducing, and socializing affects of an event to our conscious experience of it, first immediately and then in memory. Indeed, the affect of an earlier experience is one of the threads by which memory may be most easily tugged.

Memories with strong affect can be recalled by new events that bring on the same affect even if the new event otherwise bears little resemblance to the old, but the embarrassing and even awful content of such memories makes conscious recall difficult to sustain. Novelty gives an event a higher chance of being retained in memory, but when novelty is associated with intense affect, the retained memory stands a good chance of then remaining repressed — maintained in the brain but kept from consciousness — for an indefinite time, sometimes for a lifetime. Repressed memories being absent from consciousness by definition, it is tempting to argue that the idea of their existence is simply wrong. But the evidence of both everyday life and clinical observation is that repressed memories do exist in a state that is hidden from consciousness, because they can and sometimes will come to consciousness unbidden in ways we do not enjoy.

The initial retention of only certain experiences as memories, and the repression of a fraction of them, make sense in terms of our biological origins. Like our senses, our memories are processed in ways that may not serve our purposes today but that provided a selective advantage to our ancient ancestors. Selective memory and repression both diminish the load of information that must be dealt with by the parts of the brain that deal in judgment and intentional action. Without them, details of the present and the past would overwhelm our ability to plan for the future. Additionally, and perhaps even more significantly, repression is a necessary precursor to successful deception. The ability to deceive — to lie to oneself and thereby to dissimulate convincingly to others — was no doubt as important a tool of prehuman relations as it is critical to successful social behavior in our species today.

Memories begin in earliest childhood. An infant's consciousness emerges as an activity of its developing brain as the infant deals with all the aspects of a new world, including those that generate strong, even unbearably strong, feelings. The brain of a very young infant includes a repository of such unconscious memories, gathered while a newborn's consciousness is but a buzzing blur, when it still depends entirely on its parents and before it can express its own feelings through language. These first years of life contain a paradox no infant can avoid. In all that time — and for some time after — the adults responsible for its well-being hold absolute authority over it; without them the infant is alone, hungry, and miserable, yet it cannot articulate its needs and fears.

In the normal course of events, adults will help it to grow by providing it with opportunities for independence and autonomy. Paradoxically, such demonstrations, intended as acts of nurturing and love, may be felt as terrible losses: how much better it would be to have one's every need immediately satisfied than to learn to be weaned to a bottle and a schedule. As a result, deep feelings of fear and anger are directed at parents in response to a late breast or bottle, a hug not forthcoming, or a harsh voice. In dealing with such situations, an infant's emotionally rich but inarticulate mind can reach full consciousness only by passing through an extended period of deeply felt but inarticulate emotional conflict, simultaneously hating and loving the authority on which its life depends.

Such ambivalence prefigures the awkward and painful way many of us deal with similar conflicts in adult life. We are reliving our earliest experiences when we deny that we have feelings of hatred toward someone in authority or when we convert the unacceptable love or hate we feel toward a person into the notion that the person feels this way about us — or when we do the reverse, acting as if we felt toward someone the way we wished someone felt toward us. By exercising these three survival mechanisms of the very young mind — denial, projection, and introjection — we discover how to put away the pain of our real feelings, how to show a cooler, calmer face to the world than we actually feel.

When all three mechanisms fail and a painful early memory threatens, consciousness has one further defense against full exposure to its painful affect: fantasy. We all create fantasies of control, fantasies in which we have found ways to prevent the loss of love, the loss of our body's integrity, the loss of self, or, most dramatically, the loss of both self and of all loved objects we consciously realize must accompany our own death. Unlike the repressed memories that engender them, fantasies can emerge into consciousness: as dreams when we are asleep or as daydreams when we are awake. Dreams and daydreams are conscious outcroppings, processed versions of the unconscious fantasies we create to fend off our worst memories. When a daydream is not sufficient to contain an inexpressible wish, the wish may also bring about specific behaviors — obsessions — whose purpose is to fulfill it. Obsessions may be trivial — the need, for example, to wear a certain article of clothing at special times — but they are vested with enormous emotional weight.

The interaction between repressed memory and conscious perception is complicated by the way bits of unconscious memories break off and float up into consciousness as dreams, daydreams, and obsessions. The mind of a person who is fully awake, able to easily recall some pleasant moments from the past and equally good at not recalling a host of other, particularly unpleasant memories but unable to prevent them from surfacing as dreams, is experiencing many different internal times at once. Like certain genes that are expressed early in one tissue, later in another, and never at all in the germ line's frozen developmental clock, at least some of our earliest memories seem quite free from the constraints of objective time, both in their unconscious storage and in the way fantasies and daydreams recall them to consciousness.

About a century ago, a close examination of the ambiguities of time in the content of dreams and daydreams led the Viennese physician and experimental psychologist Sigmund Freud to the clinical methodology he called psychoanalysis. From his clinical observations, he came up with a series of models of the mind that included — for the first time — an unconscious component to all mental functions, including the rational ones that seemed least likely to have any relation to the unremembered past. The strategies of psychoanalysis subsequently devised through trial and error by Freud and his followers have acquired a certain mystical patina. Actually, psychoanalysis is rather straightforward.

It is based on the clinical observations that talking and listening carefully to a person's unguarded ramblings can help him to safely and reproducibly bring painful and embarrassing memories out of repression into consciousness; as careful and reflective conversation bring these memories to consciousness, it also uncovers the hidden emotional connections between current and past difficulties. The purpose of this exercise in memory recall is also straightforward in clinical terms: to help a person to learn how to release the past's control of present emotions, actions, and beliefs. Once the underlying emotional connection of the past with the present is understood, the emotional content of the current difficulty — now understood in terms of earlier events — can in many cases be brought under conscious control.

Recalled dreams are especially valuable in the analytic conversation. Without experimental data to the contrary, one might reasonably discount dreams as a form of useless and meaningless noise in the brain, perhaps the accidentally remembered residue of a nightly cleaning out of the clogged up memory stores. This is not so in psychoanalysis: once the emotional affect of an event is understood to be registered in memory, and once repression is understood to be a directed action by the brain to keep certain affect-rich memories from consciousness, dreams take on a new importance. Rather than being random noise, a dream is understood as a meaningful, condensed outcropping of otherwise inaccessible, unconscious memories, and much of analytical conversation centers on attempts to understand dreams in these terms.

Psychoanalysis reconfigured the meaning of childhood memory for all time. In the analytic model, an infant grows into full consciousness as it learns to balance perceptual information from the outside world with remembered emotional affects and conditioned habits. Full self-awareness requires merging memories with each new experience. Some memories reach consciousness easily, others — particularly of experiences or fantasies too painful, embarrassing, or threatening to consciously bear — are either repressed and remain unconscious or they reach consciousness in masked ways that lead to otherwise inexplicable behaviors.

The notion that a person's destructive, self-defeating behaviors and disturbing dreams may be conscious manifestations of otherwise repressed and unconscious impulses gave childhood itself an altogether new and somewhat ominous aspect. Many turned away from psychoanalysis deeply offended, and some still do. Disconcerting though it may be to them, and to anyone else who still dreams of an innocent childhood, the psychoanalytic narrative of the mind has withstood almost a century of scrutiny, and it remains a viable way to reach a deeper understanding of one's behavior as well as a clinically useful tool to understand, predict, and help to alleviate various self-destructive behaviors.

Psychoanalysis spent its own childhood in fin de siècle Vienna, and memories of that time and place permeate the model. Its map of the mind, with its conscious and unconscious portions and its border-crossing filters, began as a clinical protocol that depended on the oddities of the analytic conversation — slips, pauses, free associations, and descriptions of daydreams and nightmares. Some of its earliest presumptions — that a girl is little different from a boy without a penis, for instance — are likely to be based on some unexamined repressions of middle-aged, middle-class men of the late nineteenth century. Despite these self-referential flaws, Freud and the other early analysts drew on a solid nineteenth-century knowledge of the brain's anatomy in mapping the realms of conscious and unconscious thought, and the boundary between them, as the products of an inner trinity of contesting, unconscious mental states: the id, the superego, and the ego. The unconscious ego is the part of the Freudian psyche whose conscious manifestation is a grown person's sense of himself or herself; the unconscious id is the reservoir of all motivation, however irrational; and the unconscious superego is the memory of idealized authority, setting the standard of allowable thought and behavior.

Unconscious, repressed memories and their conscious manifestations — fantasies and daydreams — are major bridges between the sequential time of the outer brain's imagination and the inner brain's timeless, repressed memories. In his 1908 essay "Creative Writers and Day-dreaming," Freud described how fantasies and daydreams bridge the different versions of internal time:

> The relation of fantasy to time is in general very important. . . . Mental work is linked to some current impression, some provoking occasion in the present which has been able to arouse one of the subject's major wishes. From there [mental work] harks back to a memory of an earlier experience (usually an infantile one) in which this wish was fulfilled; and it now creates a situation relating to the future which represents a fulfillment of the wish. What [mental work] thus creates is a day-dream or

> fantasy, which carries about it traces of its origin from the occasion which provoked it, and from the memory. Thus past, present and future are strung together, as it were, on the thread of the wish that runs through them.

The Freudian unconscious of ego, id, and superego do not map completely to current diagrams of the functional anatomy of the brain, but unconscious matters of hunger, sexual desire, aggression, and fear occupy portions of the inner brain, while outer, cortical regions of the brain — especially the cortical regions behind the forehead — deal with conscious ego-like matters of subjective thought, abstraction, language, and planning. The unconscious superego's world of values, rules, standards, goals, rewards, and punishments is least centered.[1]

The first data systematically demonstrating that the unconscious repression of difficult memories was an aspect of normal brain function came from studies of survivors of head wounds. Working to understand and help brain-damaged soldiers and civilians, the Russian psychiatrist Alexandr Luria was able to partially align the analytic model of the mind with the anatomy of the brain. His most dramatic conclusion was that the normal brain was indeed functionally as well as anatomically divided into inner and outer parts. The centers of the inner brain were concerned with unconscious processes, affects, and memories; the centers of the outer brain carried out abstract conscious thought, perception of the outer world, directed action, and judgment; and at their boundary, a set of centers called the limbic systems carried out the balancing acts of bringing together the past and the present.

The inner brain described by Luria comprises the brain stem, cerebellum, thalamus, and hypothalamus, with its hormone-secreting pituitary gland. These lie between the nose and the two bumps at the back of the skull that mark the entrance of the spinal cord. Sitting at the top of the spinal cord, the inner brain receives neural information from all parts of the body and sends back chemical and electrical signals that maintain the body's posture and allow us to carry out organized, directed movements. These signals also generate the sensations of sleepiness, wakefulness, hunger, satiety, and thirst. In short, the inner brain is responsible for a person's general level of arousal; none of its neural and hormonal operations require conscious intervention.

The outer brain is made up of the parts we have already seen at work in the conscious processing of signals from eyes and nose: two wrinkled gray hemispheres of gray cortex wrapped around a mass of white cables. This white matter joins various regions of the cortex to one another, to the limbic systems just beneath, and, through them, to the inner brain. The nerves that carry the forty-cycle-per-second synchronizing hum from the thalamus to the cortex form one of these circuits from cortex to inner brain. The inner undersurface of the outer brain intimately cups the outer surfaces of the limbic systems, which also form a border — a limbus — around the inner brain. Limbic centers store our memories and generate erotic, fearful, and combative emotional states.

Unconscious memories are organized and held in the limbic region called the hippocampus; neural threads weave the hippocampus and its memories into other limbic systems and both inner and outer brains so that in the absence of repression, the emotional affect of every event and every memory can be made available to all parts of the brain. Surgery that stimulates limbic regions will generate a luminous recovery of old memories, a rich hallucinatory repertoire, and a constellation of vivid dreams.

The limbic centers are organized according to the kind of emotional affect they contribute to a momentary experience as well as any memory of it. Different emotional affects radiate to the rest of the brain from different limbic centers; Luria discovered this when he found that everyone with a lesion in one of these centers suffered the same change of personality. The clinical evidence for a region dedicated to different emotions lying at the boundary of the inner and outer brain was so redolent of the phrenologists' discredited skull models that it was some time before Luria's observations were confirmed by scientists working on drugs designed to alter a person's overall emotional response to their experience. Many of these drugs bind tightly to the nerve cells of a single, specific limbic region and to nowhere else in the brain. The limbic pleasure center was rediscovered as the major binding site for opioids, while another limbic center is the binding site for the type of antipsychotic drugs that have the side effect of reducing a person's interest in the world.

At its front end, the limbic system's centers are associated with the positive affects of pleasure, lust, and expectation. The centers associated with the three major unpleasant affects — rage, fear, and separation distress — are found in that order running toward the back and bottom of the limbic region. The limbic centers of fear and anger are closely linked through the amygdala; damaging the amygdala will generate a condition of unremitting terror and rage. The rearmost limbic center, responsible for affects associated with social interaction, is anatomically closest to the outer brain's olfactory bulb.[2]

Luria's studies confirmed that dreams are a conscious expression of memory in the absence of new conscious perceptions and that the inner and outer brains are both involved in a dream's content and expression. Different inner brain and outer brain injuries each generated differences in the way a person dreamed. For many of these brain damage syndromes, the victims also showed a specific, altered conscious state that closely resembled the dream state. Apparently, the normal function of many regions of the brain is the same in both the dreaming and the wakened mind, as damage to various parts of the outer brain shows that dreams use those parts, even though sleep keeps them free of all external sensory input.

Wounds in the white matter that lies under the front or middle regions of the cortex leave a person in a uniquely miserable condition: unable to either dream or even rest. Sleep requires that the sensory and motor portions of the middle section of the cortex be shut down, and the signals that do this traverse this region of the white matter on their way from the inner to outer brain. After such wounds, the outer brain cannot be made to shut out the world. Damage to various regions of the cortex itself, on the other hand, may also leave a person unable to dream but still able to sleep. When awake, some of these regions of the cortex process abstract symbols such as directions and rules of grammar, so it is likely that the abstract content of a dream comes from these regions of the outer brain as well.

A region of the right side of the cortex provides a sense of body location and orientation. When awake, patients with damage in this zone are terribly crippled in their sense of where and who they are; when they are asleep they cannot dream, showing that this region of the outer brain is also needed for both sleeping and wakened states of consciousness. The visual systems at the back of the cortex provide pictures for dreams, but these pictures are no more essential for dreaming than vision is for awakened consciousness. When the visual systems at the back of the cortex are damaged, a person may suffer a range of visual defects, including blindness and disorders of visual imagery of the sort Oliver Sacks has made famous, but dreams will nevertheless go on, albeit without any pictures.

The dreams of patients who have lost the region at the bottom of the visual cortex — the part of the outer brain that lies closest to the arousal center of the limbic system — shows how similar dreams can be to conscious thought. In plain sleep, our bodies are as calm and unaroused as they can be in what amounts to a reversible coma. Before we can dream, the part of the limbic system that regulates arousal must be activated. Luria found that people who lack this limbic center suffer when awake from a loss of emotions and an absence of fantasy and daydreaming. When asleep, these people cannot dream. Irritating this limbic region in an unregulated way, as in certain cases of epilepsy, heightens arousal, and all dreams become nightmarish.

The repressive capacity of the mind — its ability to prevent certain memories and fantasies from reaching consciousness — and its ability to let these memories emerge in the form of fantasies and daydreams have an anatomical correlate in the limbic systems where the two brains and their worlds meet. Damage to one particular place on the boundary between the inner and outer brains, where the most frontal of the limbic centers meets the most internal portion of the frontal zones of the cortex, has a spectacular effect on the place of dreams in a person's life: victims become unable to distinguish their dreams from reality. These lesions leave people in a permanent dream world, unable to tell whether what they see and hear is happening in the outside world or in their imagination. They suffer from "a constellation of vivacious dreams, hallucinations, confabulation and a breakdown of the distinction between thought and reality." In unaffected people, this portion of the brain must be constantly choosing among fantasy, unconscious memory, and current reality.[3]

Each part of the brain contributes to the one inner voice of consciousness. Even the conscious act of learning from an event — the minimal unit of scientific observation — is the sum of at least four different kinds of neural activity taking place simultaneously in the two brains and in their shared limbic boundary. At the conscious level, the right cortical hemisphere of the outer brain internalizes the sensory experience of the event in terms of self-definition — what does this mean to me? — while the left cortical hemisphere retains the event cognitively, in language, as a set of facts and observations. Simultaneously, the forty-cycle thalamic clocks enable the limbic system to attach an emotional affect associated with the event. The hippocampal memory retains a trace of both the event and the affect, while the hypothalamic regions of the inner brain generate their unconscious responses to the event. Of these various expressions of the brain's functional anatomy, someone doing mental work — a scientist analyzing her data, for instance — is consciously aware of only the first two.

The emotional affects and memories of the past will not be part of a scientist's conscious experience, but because they are registered as changes in brain circuitry, they necessarily will be part of each act of observing and understanding the natural world. Scientists may insist that no aspect of nature is hidden from them, but inevitably their own nature — the conscious manifestations of their unconscious fears and needs — shapes the questions they ask of nature and thus what they can discover about the body and the mind. With that in mind (so to speak), let us apply this empirical, clinical model of the relation between conscious thought and unconscious memory to the question of how the agenda of a science may keep unwanted unconscious memories from emerging to trouble the consciousness of its scientists.

The conscious part of science begins with an act of faith, the ancient Greek belief that the natural world works by mechanisms that we can understand, even though they may initially be hidden from view. Today, as in Democritus's time, science works within the Greek belief that despite the smallness of atoms, the largeness of the cosmos, the rapidity of atomic transmutation or chemical catalysis, and the imperceptible slowness of evolutionary change, the underlying reality of any aspect of nature will be consistent, understandable, and therefore knowable. Everyone who plays the game of science must come to it infused with the belief that the way the natural world works can be understood to any degree of detail by sufficiently clever experimental manipulation.

Since nature is clearly silent and uninterested in a scientist's curious faith, the first step toward understanding is for the scientist to come up with a hypothesis to explain how some natural phenomenon works. The hypothesis is then tested through experiments that compare its predictions to the actual behavior of nature. Good experiments must often use elaborate machinery, so science can be expensive. But a good experiment need not be complicated because it is never simply a set of measurements; it is a test of the usefulness of a figment of the imagination and a moment of risk and drama.

If experiments confirm a hypothesis — for why a ball bounces, a cell dies, the moon turns, or a muscle contracts — then the scientist must expand the range of tests to determine whether the hypothesis explains a little or a lot. As in backgammon, the stakes in science always go up; the game is never more risky than when a hypothesis proven right in a small corner of the natural world is tested in a bigger one. Each successive confirmation carries with it the obligation to push a hypothesis into ever-larger realms of nature by more extensive and subtler experiments. When — usually sooner than later — careful experimentation confounds a hypothesis, it has reached the limits of its usefulness, and it must be redrawn or withdrawn. One may think that a hypothesis that explained even a little would be treasured and preserved. But once a hypothesis has been bounded by contradiction, the faith of science demands that it be altered or entirely replaced so that the task of understanding nature more fully may go on. Solely on the conscious level, science is thus reduced to a mixture of ritual and game, complete with a game's obedience to its own rules, austere unworldliness, and willful naïveté.[4]

The conscious part of science is what most scientists would insist is all there is to science: an agenda for understanding nature. However, based on what we know of some minds, we can expect that the minds of scientists and therefore perhaps even the mind of science — the communality of experience and motivation shared by most scientists — have both conscious and unconscious parts. Just as the conscious part of science is shaped by the set of simple and universal rules that govern the conscious activity of all scientists, engaging and pooling the efforts of many different people's conscious minds, the unconscious parts of the mind of science — in particular, the sciences that serve medicine — would be expected to emerge as fantasies and obsessions shared by scientists in these fields.

While the notion that scientists may share their unconscious fears and conscious fantasies, dreams, and myths may seem disingenuous, meaningless, or just plain silly, recall that until not too long ago, many serious observers thought it was disingenuous, meaningless, or silly to imagine that an individual brain might contain — within its biological functions — any individual mind at all. The early behaviorist assumption that the mind is an illusory, ineffable byproduct of the brain's mindless application of instinctive rules had to be set aside in light of what we now know about the brain's functional anatomy. It is time to follow up on that conclusion, to set aside the notion that science can operate in the present moment without an unconscious component to its deliberations. Science is the product of the unconscious sources of imagination and introspection as much as it is the product of a set of rules. The emotions and memories shared by scientists in the same field are its inner voice, and there is no reason that these inner voices should not be dealing with the same unconscious, repressed memories as do any of a field's practitioners. The question is not whether but how the unconscious aspects of science, refracted in maturity through its methods, resurface in ways that deflect the course of science itself.

Taking into account what we now know about the mind's

operations in the brain, we can predict that shared fantasies of science are likely to be built from early memories of scientists, especially memories storing very strong negative or positive limbic affects and ordinarily kept from consciousness by repression. Though science may seem at first remote from unconscious memory or conscious fantasy and obsessive behavior, it remains a human enterprise, and the fantasies of infancy are likely to be the same, whatever a person's later career. When negative affects are dealt with in the same way by a group of people linked by language and culture, their shared fantasies can crystallize into a core of collective myth.

Every time biomedical scientists look at a piece of the human body through the lens of science, the lens becomes a mirror. What they see in it is at once familiar and completely strange. The human mind and body, but especially the mind and body of the scientist, become uncanny; the German word *unheimlich* best conveys the way each becomes more strange as it becomes better understood. This uncanny element of the life sciences derives from the fact that we can neither fully accept nor consciously and rationally even understand our own death. In Freud's words, "Biology has not yet been able to decide whether death is the inevitable fate of every living thing or whether it is only a regular but yet perhaps avoidable event in life." In its confrontations with mortality, medical science places itself in an ambivalent position toward an authority that is no longer anyone's own parent but the parent of us all, nature itself.

Nature makes us mortal; surely that affects the behavior of every thinking person. For the scientist, what better way to reduce the feared figure of our own mortality than to make it our experimental material? This is a modern version of the original Greek notion of science, born in a world that did not distinguish between science and religion. The underlying myth of science concerns one of their immortal but otherwise altogether human gods. Asklepios, the demigod of medicine, was the son of the immortal god Apollo and a human princess named Coronis. The centaur Chiron — a physician of consummate medical skill who also happened to be an early human-horse recombinant hybrid — taught Asklepios the arts of medicine, and Asklepios became so skilled at healing that he was able to resurrect the dead. Hades, the immortal god who ruled the underworld, complained to Zeus, the father of the gods, that he feared the loss of future subjects if humans were no longer to die. Zeus's response — he killed Asklepios with a thunderbolt — explains our present mortality and leaves us with the fantasy that by rediscovering the skills of Chiron and Asklepios, we may yet one day escape death.

The rest of the myth as it has come down from the Greeks tells us that Apollo, the immortal god of song and light, took such offense at this act of Zeus that he slew the Cyclopes, the makers of Zeus's thunderbolts, in revenge. Revenge, however, came too late to help poor mortals then or now. This myth and the hope it expresses have survived for thousands of years longer than any of the gods it describes. Despite all the rewards of scientific understanding, Apollo — science itself — has not yet overcome mortality. But we all vest the same hope in Asklepios — consciously or unconsciously — just as the Greeks did each time we visit the doctor.[5]

The original Greek myth is alive not only in the minds of patients; it also lives in the minds of many doctors and scientists. Sometimes a great scientist will let the dream of Asklepios surface, allowing it to peek out from behind that other ancient Greek mask, the rationality of science. For instance, in his autobiography, the French Nobel laureate François Jacob is ostensibly discussing what it feels like to carry out a series of experiments, and the American Nobel laureate Arthur Kornberg is writing an editorial explaining why the scientific endeavor is unique among human activities, when both emerge with unexpected confessions of faith:[6]

> JACOB: And with this idea that the essence of things, both permanent and hidden, was suddenly unveiled, I felt emancipated from the laws of time. More than ever, research seemed to be identified with human nature. To express its appetite, its desire to live. It was by far the best means found by man to face the chaos of the universe. To triumph over death!
>
> KORNBERG: The ultimate scientific languages used to report results are international, tolerate no dialects, and remain valid for all of time. . . . Science not only enables the scientist to contribute to the progress of grand enterprises, but also offers an endless frontier for the exploration of nature.

Only faith or obsession — if they are not the same — can expect a method for observing nature to give a vision of endlessness or of triumph over death. Hyperbole like Jacob's may be intended or read as metaphysical metaphor, but the underlying fantasy remains clearly expressed: omnipotence of thought will bring immortality. This notion does not stand up to rational analysis; that is why the conscious, operational agenda of science masks the fantasy in Kornberg's "endless frontier," the cloak of institutional immortality. But institutional immortality itself, born from the unconscious will that one's name not be scattered, is just a different version of the same fantasy, an ancient impulse not limited to the sciences.

Myths of immortality — personal or institutional — distort scientists' conscious behavior. They steer the game of science in directions that have less utility than the scientists themselves may believe but that point away from an explicit confirmation of the underlying fears that create these myths. The uncanny familiarity of death, always on the threshold of being rediscovered by the rules of science, obliges scientists unconsciously to subvert the rules of their game, turning away from some of their most important discoveries.

The denial of the fear of nature's terrible power of mortality, the projection of the suppressed wish not to die into a vision of nature as capable of bestowing immortality; these are the marks of a masked unconscious creating a biomedical science at war with its own stated purposes. When scientists say "give us your bodies, and we will cure you," they have found a way to deal with an otherwise unbearable ambivalence toward their own experimentally vulnerable existence. They protect themselves, not necessarily by curing anyone, but by gaining control of someone else's body if not their own. Freud recognized this role of medical science as the one "higher superstition" he himself believed in: "My own superstition has its roots in suppressed ambition (immortality) and in my case takes the place of anxiety about death which springs from the normal uncertainty of life."

In its disciplined way of looking at the natural world, science requires its practitioners to act as if they were observers, not participants. The first and last scientific instrument, the one that must be used in every experiment, is the scientist's brain; scientists who choose the human body and mind as their playing field cannot fully meet this requirement without dislodging themselves from their own bodies and minds. The strain of trying to meet a standard of dispassionate curiosity without flying into pieces imposes an irrational gap between the student of the brain and the brain of the student, between the scientist and his or her body and mind. To deal with the emergence of this intolerable thought, medical scientists have created the myth that their instruments and procedures somehow free them from the boundaries of their minds and bodies. This is the myth of absolute rational control of the physician and scientist over their material, the notion that the metaphor of scientist as sculptor will not break down even when the sculptor and the sculpture are one and the same. This myth may work to keep thoughts repressed, but at the cost of requiring the belief in an invented, institutional immortality based on the dry fact of precedence.

Every discovery must have at least one discoverer, and many have more than one. As competing sculptors may race to clear the excess stone from blocks of marble to reveal their different visions of what lay hidden inside, competing scientists clear away layers of plausible models, racing to uncover a demonstrably accurate schematic explanation of a part of the natural world's inner workings. Consciously and conscientiously followed, these rules work; they permit at least some scientists to uncover the mechanisms and structures of the natural world, and they do permit a few to win the game.

The dream of winning takes on an obsessive quality in the medical sciences, where the subject of scientific study is the mind and body and the reality of mortality becomes unavoidable. The result is an obsessive hope: that a big enough win in the game of science will confer a form of immortality on the winner. Discoveries that set the agenda for the future work of a large group of other scientists do this after a fashion, permanently associating a scientist's name with an aspect of nature. Think of the Freudian slip and the Watson-Crick model of DNA. Players in a game that can confer even this sort of immortality — however rarely — cannot be playing only for conscious stakes. In the medical sciences, the belief in winning immortality of this sort can become problematic when it supports the denial of an unpleasant biological reality, especially when that reality emerges from precedent-setting discoveries themselves. It is not that science and medicine wish to avoid finding cures. It is that they are too strongly motivated by an irrational, unconscious need to cure death to be fully motivated by the lesser task of preventing and curing disease simply to put off the inevitable end of their patients' lives and, by extension, their own.

No scientist — nor any science, no matter how rich or creative — has time to look at all the data that all possible experiments may generate; it would be a form of madness as well for any single scientist to examine all the data — all the memories and feelings and perceptions — of his own brain, bypassing the filter that keeps much of it unconscious. Instead, every science chooses selectively all the time, and with each choice some data are precisely not gathered, let alone examined. Choices are necessary, and it is at the moment when choices are made that the scientific method departs from the wholly conscious tool of scientific experimentation and enters the human world in which all choices are made in a personal and social historical context, replete with emotional affects and barely remembered feelings.

There is a way for the life sciences to end their denial of their own unconscious, freeing it from the obfuscations and inefficiencies it creates today out of its own fantasies. An enlightened medical science would acknowledge that there are limits to conscious thought and to life itself that cannot be transcended by any rational agenda. It would then be able to stop making promises that it cannot keep, whether to itself or to the rest of us who pay its way. Having next acknowledged the unconscious memories of its practitioners and the shared fantasies they have generated, it would then be ready to find ways to diminish the influence of these fantasies on its conscious agendas.

The three major causes of premature death and avoidable suffering in the human species are infectious diseases, malignancies, and the consequences of aging. Scientists and doctors who work to understand these causes do so for many motives, conscious and unconscious. When one realizes that an infectious disease is in fact the invasion of the body by an invisible army; that cancer is in fact an insurrection of the body against itself; and that aging, dying, and death are not metaphors to scientists and doctors but impenetrable barriers to their success, it is not hard to guess at the unconscious wellsprings of fantasy in the medical sciences. In the next chapters, we look at these three lines of scientific research and ask two questions. First, are their priorities in fact distorted by the denial of an unconscious fear? And second, what would be different about science if these fears were acknowledged, not denied? Let us look first at why infectious diseases are so hard to wipe out, what is being done to ameliorate their devastating effects, what unconscious fears they evoke, and how these fears may be inhibiting our ability to come to terms with them.

4

The Fear of Invasion

The rationality of life is identical to the rationality of that which threatens life.

— Michel Foucault, *Birth of the Clinic,*
quoted in W. Biddle, *Germs*

Science cannot utter a single word about an individual molecule, thing or creature in so far as it is an individual but only in so far as it is like other creatures.

— Walker Percy, *The Message in the Bottle*

THOUGH OUR MINDS WORK as if each of us were a single, autonomous entity, the body is permeated with other, invisible lives. Most of these intimate neighbors respect the skin as a boundary between themselves and our cells, but some do not. Dry on the outside and moist within, the continuous coat of dead skin cells that separates the body from the world is constantly breached by microbes, some of whose lives depend on getting inside one of the live cells beneath it. They usually do not get very far beneath the skin before they are met by cells from the body's immune system. As a tissue, the immune system is as big as the brain and spinal cord combined. It is less noticeable than the brain only because it is not in one place but spread under the skin, the better to find and kill invading microbes before they can establish a beachhead. If the brain is the center of the neural networks, the gut comes closest to being the center of the immune system: more than half of the immune system's cells lie just beneath the moist skin lining the tube that runs from lips to anus.

The immune system is similar to the nervous system in many ways: it is about as large, it is about as complicated, and membrane-bound receptors on some of its cells give the system as a whole a comparable capacity to sense and respond to changes in the body's situation. The main differences are that the immune system is designed to sense changes in the body, while the nervous system senses changes outside it, and that the immune system is not linked to the brain's cortical processes we call consciousness, so its operations are for the most part odorless, tasteless, and invisible — in a word, insensible. Because the brain is not directly wired to the immune system, we cannot notice whether or not a microbe has gotten past the skin. We sense an invading microbe only when the ensuing interaction between microbe and immune cell generates the symptoms of an infectious disease.

The smallest and simplest of infectious microbes are the viruses. A virus can cause the symptoms of a disease only after it has found its way into one of the cells that make up a person's body and begun to reproduce. Viruses with fewer than a dozen genes or more than a hundred are rare. Bacteria are many times more complicated and larger than viruses, though still far smaller than any of our cells. With thousands of genes on their chromosomes, they are complex enough to live on their own. Most kinds of bacteria are finicky, growing only under narrowly defined conditions. For thousands of bacterial species, these conditions are best met somewhere in our bodies; for the most fastidious, nothing less than the inside of one of our cells will do. The most complex microbes are the protists: single large cells with the same architecture as one of our own cells, complete with a nucleus holding chromosomes that contain tens of thousands of genes. Like the bacteria that find our bodies hospitable, many protists grow best — or only — inside one of our cells.

Regardless of the particular symptoms, each case of an infectious disease is a Darwinian struggle for survival between microbes and the cells of our immune system. Each infection can have only one of three outcomes: the microbe dies, the person dies, or the microbes and the immune system reach a truce. Microbes and immune system cells bring the same weapons to this struggle for survival: evanescence and genetic malleability. Just as every brain changes its synaptic connections in response to life's experiences, every immune system is constantly retuning itself. When a person recovers from an infection, it is because the infecting microbe has selectively stimulated the growth of cells from the immune system with the capacity to recognize it, neutralize it, and then remember to do so if it reappears. The persistence of a very low dose of the infectious microbe is a goad; in cases where there is complete microbial obliteration, there is also a loss of immune memory and the risk of serious disease the second time around.

Microbes have different ways to get past our defenses as they move in and out but only one way to grow: simple division. One microbe can become two, and two four, in an explosive chain reaction of growth that can overwhelm an infected cell in a few minutes and an infected body overnight. Like our own germ cells, most microbial offspring are destined to die without leaving any trace. Just a few need to survive and infect another person in order to keep the microbial species going. All microbial strategies for survival in the body are built on the same basic plan of infection and reinfection: seed one body, multiply in it, seed the next.

Interludes of reinfection are as essential to microbial survival as children are to the survival of our species. All successful microbes have found ways to negotiate the leap from one person to the next, that is, to be contagious. Contagion makes even the most dangerous microbes depend on our behavior, putting them in an unavoidably precarious ecological situation. In the few weeks of a flu infection, for instance, hundreds of generations and billions of viruses will be born inside the cells lining a person's bronchial tubes and nasal passages. All but a few of the virus particles will die in the body or be sneezed out to dry to death in an inhospitable Kleenex. But so long as a single virus makes it from one of the victim's cells into another human cell — the victim's or anyone else's — before dying, the population of viruses will survive and persist through time.

Like songbirds in a suburb, infectious microbes are a metapopulation, living on islands — backyards on the one hand, sick people on the other — separated by hostile territories in which they cannot survive for long. Microbial metapopulations no more demand the suffering or death of an infected person than avian ones demand the loss of woodland. Birds maintain metapopulations by migratory flights before nesting, and microbes do the same. Just as songbirds may disappear from one neighborhood as they migrate to another, microbes die off when a person either gets well or succumbs to the disease. The symptoms of illness induced by a microbe often occur simply because they help it to reach another host. For example, cholera bacteria do not benefit because the diarrhea they cause leads to the death of their host but because the diarrhea spreads the bacteria around so quickly and widely that some have a good chance of reaching a new intestinal environment, even if the infected person dies in only a few days. Other microbes follow a waiting strategy, walling themselves off from a hostile world as spores until they can get inside a susceptible body; for example, anthrax spores can remain viable for decades in dry soil and then germinate to cause a serious lung disease when a few are finally inhaled.

Flying creatures — birds and insects — establish the most wide-ranging metapopulations, passing over large inhospitable areas to establish small, temporary pockets of fertile growth. Some microbes have adapted to a flying host's tissues as well as our own in order to expand their capacity to establish their own metapopulations in people. A bird or an insect can bring a microbe from one person to another, when by themselves few pairs of people would make sufficiently intimate contact to accomplish the microbial handoff. Insect-borne microbes can create great misery over a wide area because their insect host can transport them miles away to their next human host even as their current victim lies immobilized by illness.

The malaria parasite, for example, grows so well in red blood cells that even a mosquito bite's worth of blood from an infected person will be loaded with parasites ready to be transferred to the mosquito's next victim. But without the mosquito, these parasites are all marooned in the sick person's body, unable to reach a fresh supply of red blood cells. The two-host strategy works for bacteria and viruses as well as protists: epidemics of mosquito-borne yellow fever, louse-borne typhus, and flea-borne plague are examples of microbial metapopulations moving very rapidly from person to person inside their second hosts.

Whatever their strategies for contagion, all successful microbes must also elude the immune system. Immune cells called macrophages will ingest any microbes they find, informing other immune cells of the invader so that they can zero in on infected tissues, killing any cells already harboring a nest of growing microbes and marking the infected area with chemicals called cytokines for later rescreening. The power of the immune system is unleashed by microbial growth; it cannot be exerted if a microbe gets into a cell and simply sits there. Recovery from microbial infection is rarely a total victory of the immune system over a microbe. More commonly, it is the cold peace of mutual coexistence between an armed immune system and a genetically unstable, ever-shifting population of temporarily domesticated microbes.

In establishing this compromise with infectious agents — grow and I'll kill you, sit around and I'll leave you alone — the human immune system has encouraged the differential survival of the slow-growing, well-hidden variants of many contagious microbes. Sometimes the compromise leads to true domestication, but if a microbe comes out of hiding and begins to grow, it can do a lot of damage. The microbe responsible for tuberculosis, for instance, audaciously sets up house inside the very immune cell that catches it as it enters the lining of the lung. There it lives in suspended animation, surviving for decades without causing symptoms, just in case it has the chance to flower into rapid growth — causing the immediately devastating, acute symptoms of tuberculosis — should the host's immune system for any reason begin to fail.[1]

Microbial genetic variation coupled to changes in human behavior can lead to entirely new means of contagion. A few centuries ago the plague bacterium mutated into a form that could live well in a person's lungs, moving from a sick person to a well one by a cough, as if it were tuberculosis or the flu. This variant, called pneumonic plague, no longer needed to pass through a flea to get from one person to the next. In the new, unsanitary cities of medieval Europe, pneumonic plague killed about one person in three before it died out for want of new, immunologically vulnerable hosts.

Microbes continue to adapt themselves to our changing habits. Each new infectious disease is a new perturbation in the equilibrium between our species and the microbial world. Our behavior can shift the equilibrium, for better or worse. Travel, for instance, shifts the equilibrium in favor of a new microbe by giving it new chances of finding its way into a body with a naive immune system; every new way we invent for traveling about the planet makes us more vulnerable to emergent diseases. When plague emerged in Europe about seven hundred years ago and killed about a third of the population, it had traveled at about three miles a day from Asia, and the trip took about a thousand years.

When European explorers brought smallpox and measles to the New World five hundred years ago, the trip took a year by ship. Cholera went by boat from Asia to Peru in a few weeks. A decade ago, an emergent virus called Ebola flew from Africa to a laboratory in Virginia in a few hours in an inadvertently infected monkey being shipped to a research laboratory. Insect vectors travel with us as well: a survey of sixty-seven airlines arriving at Heathrow Airport in London a few years ago found healthy insects — including mosquitoes — in the wheel wells of about one airplane in twelve.

All people are potential vectors, carrying new pathogens in and on their bodies as well as in their luggage. Because the environments and the immune memories of people from different places are different, each is likely to carry a different spectrum of pathogens. As people from different parts of the planet mingle in airports and cities, microbes that live on the skin and all variety of intestinal flora get a chance to jump around and to test a range of human environments they would never see in the natural course of events. In 1800 fewer than two percent of the world's population lived in cities; by the year 2000 — the population as a whole having increased by more than a hundredfold in two centuries — more than half of all people will live in cities, and five hundred or more of those cities will each house more than a million people. Besides accelerated growth, the other hallmark of the twentieth century — total, mechanized war — has provided particularly attractive opportunities for microbial variation. The more people, the more habitat islands for any human pathogen; the more complete a war, the less medical interference to their spread.

In peacetime, the growth of cities and the widespread conversion of forest edge to suburb bring many people into contact for the first time with insects and birds bearing novel microbes: Lyme disease is an example of deer, mice, ticks, a bacterium, and people all meeting together for the first time. At the same time, one major large-scale consequence of our numbers and our reliance on fossil fuels for transportation and industry — a global increase in average annual temperature — has brought tropical insect vectors into temperate climates, allowing mosquitoes to bring malaria to large numbers of new, uninfected people.

Our own chromosomes are the most elegant hiding place of all for the genes of any microbe clever enough to insert its DNA into one of them. If even a piece of a microbe's DNA can find its way into a chromosome in one cell of the body, it can then piggyback its own copying onto the copying of the cell's DNA and so always be present in all the descendants of that cell. HIV — the small virus responsible for the symptoms of AIDS — does this quite well: it tricks an immune cell into seeing viral genes as its own by inserting itself into chromosomal DNA. From then on the virus has no need to do anything; every time the cell divides, it makes new copies of the virus's genome. The most clever of all hiding places is the DNA of a cell that makes sperm or egg cells: from such an eyrie, microbial genes can be assured of survival for as many human generations as the human family of their choice continues to propagate successfully. They never have to spring out and confront any descendant's immune system. Once in the germ line, a microbial sequence need never function at all in order to survive indefinitely. The result of such luxury is degeneracy: the human genome is riddled with nonfunctional sequences that had once been potentially virulent microbial genomes until random mutation destroyed their meaning.

Vaccination — the intentional, preemptive infection of a healthy person with a benign, domesticated version of an infective agent — prepares the immune system for later, more dangerous infection. Vaccines were developed more than two centuries ago in response to the smallpox virus, well before the discovery of the virus itself or of the immune system. Before vaccination, a runaway smallpox infection was a miserable experience, with disfiguring scars on the skin as the mildest outcome and death as a serious possibility.

The smallpox virus grows so rapidly beneath the skin that the cells it kills merge into blisters filled with clear, viral soup. The fragility of the blisters makes them excellent vehicles of contagion, delivering an infectious dose of virus to any breach in the skin of a person touching one of them. The blisters defined the disease: smallpox survivors would always have deep pits — eponymous pocks — at the sites of their blisters. Experiments carried out in the late 1600s — the sort of experiments we now consider to be acts of criminal irresponsibility — showed that a relatively mild case of smallpox would result from putting a tiny bit of the fluid from a victim's sore onto a scratch made in an uninfected child's skin.

The intentional administration of a potentially lethal infection — called variolation, after variola, the name of the disease at the time — did in fact work, at least some of the time. A variolated child usually would get a mild case of smallpox, with a big pock at the site of the inoculation, but then be immune to further infection, surviving any number of subsequent epidemics that might swirl through a city like London or Berlin. There were problems with the technique. Sometimes the variolation would simply kill the child; more often variolated children would themselves get well but be sufficiently infectious for a time to trail a wake of lethal infections as they recovered from their own illness. The first recorded variolation in the United States was performed by the Reverend Zebdiel Boylston on his son and two of his slaves; there is no record of the good reverend trying it out on his own body.

By making a single, subtle change in the accepted procedure of variolation, the British physician William Jenner created the first vaccine. "Vaccine" is taken from the French for "cow," because the first viral domestication was the unexpected consequence of people and cows living in close quarters. Farmers and milkmaids who drew milk from a cow often survived their first smallpox epidemic without any disease at all. Jenner knew that protection depended on their having been exposed earlier to cowpox. Cows infected by the cowpox virus would have pocks on their udders that quite resembled smallpox lesions, and as they were milked, these pocks would spread the virus to the milkers' hands. Exposed milkers would experience a cowpox infection as a mild fever and rash and thereafter bear a few, small pocks. It is one thing to notice that milkmaids are protected from smallpox by their cowpox scars; it is quite another to propose to infect someone with cowpox to see if it would protect them against smallpox as well. Oddly, the idea of infecting a person with a cow's virus bothered the very doctors who were happily infecting people with smallpox by variolation. Experimenting on people with live viruses seemed not to be a problem, but the notion of mixing living material from different species was.

In the late eighteenth century, a doctor could still do pretty much what he wanted as long as he was a British gentleman, so Jenner was free to intentionally infect his children and servants with cowpox by scratching some of the fluid from an infected udder into their arms. The uncontrolled, unethical "experiment" worked. His cowpox-infected children and servants first developed a mild set of symptoms, then one small pock at the site of the infection — the first vaccination mark. Thereafter they could go through epidemic after epidemic of smallpox without any ill effects while their unvaccinated friends and relations sickened and even died.

Jenner's vaccinations depended on a sick cow, a pharmacopoeia too fraught with uncontrolled variables even for that time. His solution for maintaining the vaccination material was as novel and as radically inconsiderate as his initial decision to try the cowpox on his children: he always made sure to have a freshly vaccinated person on hand so that he could vaccinate new people with the material from the recovering person's vaccination sore. To bring his technique to the United States and India, he used relays of immigrant children, vaccinated from one another on the journey. The pox-virus odyssey ended in the mid-1800s, when someone infected the udder of a cow with fluid from a newly vaccinated person's sore and thereafter used the cow's sores as a source of vaccination material. That trip — from cow to human and back to cow — created the vaccinia virus. In its travels, vaccinia has accumulated a very large set of genetic differences from both smallpox and cowpox viruses, but it was never a wholly safe agent to administer. Preparations differed in potency, and the intentional infection sometimes raged into a fairly good imitation of smallpox itself.

Despite this checkered past, vaccination using stable, standardized preparations of vaccinia eradicated smallpox by the early 1980s. Killing smallpox took a hundred million free vaccinations as well as twenty years of unprecedented and subsequently unmatched international cooperation. In 1967 there were a million deaths from smallpox worldwide; in 1980 there were no deaths and no cases were reported. The very last recorded death from smallpox in the West did not involve infection: in 1978, after a technician at a virus-stockpiling laboratory in England died of inadvertent infection from an archival strain being handled in the next room, the laboratory director committed suicide in his disgrace.[2]

Vaccinia remains alive and well, and it is the subject of intensive research aimed at using it as a vehicle for the introduction of single microbial genes that would convert it into a benign Trojan horse, useful for "vaccinating" a person against any number of other infectious agents. Smallpox remains in suspended animation in a few laboratories, preserved by scientists curious to know the relationship of its virulence to its DNA sequences; although the move to destroy the last stocks gains force every year, the argument that its sequence should be obtained first has so far held sway. Smallpox was conquered by the cooperation of hundreds of millions of people acting in their own interest, not by any particularly elegant basic medical science. Even with our immune systems at work inside us, we needed mass education, a willingness to come clean about the presence of the disease, and a gift from the industrial world to its poorer parts of hundreds of millions of doses of an effective if poorly understood vaccine to eliminate smallpox completely.

Medical science has not eliminated any other infectious agent. Some, like the bacterium that causes tuberculosis, the parasite that causes malaria, and the virus that causes AIDS, are currently killing vast numbers of people each year. Tuberculosis and malaria are the most prevalent microbial diseases we face today, and neither has yet been domesticated into an effective, live vaccine. Until a decade ago, antibiotic chemicals that kill microbes even as they hide inside a cell seemed to be the perfect complement to the immune system and its boosters in dealing with such bacterial diseases. By getting inside a microbe or an infected cell of the body and specifically blocking a step in the growth of a virus, bacterium, or parasite, an antibiotic can deliver a clean and complete cure, removing an offending microbial population down to its last evanescent individual. But antibiotics are also agents of natural selection, and the microbial strategy of survival through mutability and evanescence thrives on the challenge of a simple chemical. In their initial success, antibiotics destabilize the equilibrium between an immune system and a microbe, first in a favorable direction, but then sometimes tipping the balance against us. The rare microbe born at random with a mutation freeing it from the grip of an antibiotic will overgrow its dying cousins, producing no recovery but rather a new, antibiotic-resistant infection.

Although the outer wrapping of a tuberculosis bacterium is the steel-belted radial tire of microbial membranes, impenetrable by most antibiotics that work against other bacteria, about a dozen specialized drugs have been found to halt or reverse a tuberculosis infection. One, called INH, is capable by itself of stopping full-blown tubercular overgrowth; the others are less effective. INH worked so well in the 1970s — and strains resistant to it were so rare — that it was assumed that tuberculosis had been conquered. The notion that an organism with the capacity to outwit the immune system would not be able to mutate into a state of INH resistance was willfully naive. Like all other microbes, the tuberculosis bacterium turned out to be a minimalist at self-defense: a single mutation rendered it completely resistant to INH. In 1986, just as tuberculosis was beginning to surface in AIDS patients, the United States government's Centers for Disease Control and Prevention — the CDC — took a step that cannot be explained except as an example of deep and pervasive institutional denial: it ceased to require hospitals to report cases of INH-resistant tuberculosis.

The timing could not have been worse. With this decision, INH-resistant tuberculosis was given the time and chance it needed to send forth variants that would be resistant to other drugs as well. Drug-resistant tuberculosis killed many patients in public hospitals in the 1980s, especially the ones who were too sick with AIDS, too confused, or too disorganized to take their tuberculosis medicines long enough after a first bout of the disease. At the end of the 1980s, multiple-drug-resistant tuberculosis was the most common way a person with AIDS died. The CDC caught up with INH resistance by requiring that any person showing symptoms of tuberculosis take a cocktail of four drugs including INH for six to twelve months. If that mix did not make matters better after a few weeks, it was to be replaced by a second formula using the remaining seven drugs known to work against the microbe.

Given the contagious nature of full-blown tuberculosis, every one of us is at the mercy of a patient taking either of these cocktails, which must be taken for a full year after they clear up the symptoms of tuberculosis. If a patient takes the drug only until the symptoms diminish, the remission will fail: unless the drugs are given in high enough doses and for a long enough time to kill all tuberculosis bacteria, the ones that survive are all but certain to come back when the drugs are stopped and cause a new round of tuberculosis. Worse, any drugs that are stopped prematurely once are more likely to fail should they be tried a second time, because the bacterial population that survives the first round of treatment will be enriched for drug-resistant variants.

Today, doctors and public health officials in many large American cities no longer hand out drugs to tuberculosis victims. Instead, they require that patients be observed taking their medicine every day. Directly Observed Treatment, Short-Course — DOTS — is remarkably effective in assuring that a person infected with tuberculosis will follow the full regimen of drug treatment and not be the incubator for a new drug-resistant strain of the microbe. Even though it offers no further insight into molecular mechanisms of resistance, DOTS seems to protect against the tuberculosis microbe's strategy of adventitious genetic variation. It is a good example of the utility of acknowledging and working with a microbial strategy, instead of waiting for an endless stream of new antibiotics to save the day.

DOTS has not been widely used outside the inner-city tuberculosis clinic. Because such a large percentage of tuberculosis patients are people in poor nations, the wealthy nations of the world would have to spend a large amount of money to stockpile and deliver the drugs needed to carry out DOTS therapy in a complete, orderly way. So far this has not become part of the scientific agenda of the wealthy nations that do fund medical treatment in the world's poorest countries. Consequently, three million people — almost all of them in poor countries — die of DOTS-treatable tuberculosis each year; half a million of them are children.

The malaria parasite lives well in the human body through a different strategy: it is notorious for the ease with which it escapes the immune system, which recognizes the parasites by the proteins on their coat. People usually have no difficulty mounting an initial immune response to a malarial infection, but then their immune systems simply cannot keep pace with the parasite as it picks new combinations from its genes to make altogether new coat proteins. A variant parasite whose new coat allows it to slip past the immune system will always take advantage of its moment of freedom and fill the body with its progeny. This genetic malleability is more than a match for the immune system's own ability to turn on a dime, and so far it has kept us from finding an effective vaccine for malaria. It is so delicate that we cannot yet grow it in a pure culture and still must do experiments in infected animals. Some laboratories have tried purifying some of the outer proteins of the parasite and making vaccines out of synthetic fragments of them, but neither these, nor any vaccines made from isolated parts of the parasite, have shown more than marginal effects in the field.

Current research on malaria has focused on a search for effective antibiotics. Quinine — drunk in some cultures as an extract of tree bark and in others as a gin-and-tonic — is a prescientific antimalarial drug of some usefulness and the template from which most other antimalaria drugs are built. The quinine family of antibiotics all work the same way: they force the parasites in a red blood cell to choke on their own gluttony. After finishing off the red blood cell's hemoglobin, the parasite has to spit out the molecule's indigestible pit, an iron-rich compound called heme. Quinine and its synthetic derivatives chloroquine and methoquine all prevent the parasite from pushing heme out of itself into the red blood cell; eventually the piled-up heme kills the parasite.

These drugs all rapidly select for resistant strains, especially when they are not taken in sufficiently high doses for sufficiently long times. Methoquine, a second-generation variant of chloroquine, took about seventeen years to develop, but methoquine-resistant strains appeared only ten years after it was first administered to people suffering from chloroquine-resistant malaria. Malaria can be prevented by killing off the mosquito that transmits the parasite, but using insecticides for this purpose has the same side effect as any other antibiotic: insecticides like DDT once were the scourge of mosquitoes carrying malaria; today malaria is carried almost exclusively in DDT-resistant mosquitoes worldwide.[3]

Our species has had one last, desperate response to malaria: human genetic resistance. A child born in a region where malaria has been common for a long time — or born of ancestors from such a region — has a good chance of inheriting a variant hemoglobin gene that protects against malaria. If she inherits a protective variant from one parent and a more common version of the gene from her other parent, her red blood cells will be inhospitable to the malaria parasite and she will have a life-long resistance to malaria. But a child who inherits the protective variant of the hemoglobin gene from both parents will soon develop sickle-cell anemia, a serious problem resulting from a far greater variation in the structure of red blood cells. It hurts, and in some cases it can be fatal, to have red blood cells form sharp-edged stacks instead of smooth pillows and to have them jam up the tiniest blood vessels of the muscle and lung. The pain and death of sickle-cell disease and other blood disorders is the price our species has paid and will continue to pay to escape from the otherwise unstoppable combination of mosquitoes and malaria.

A parasite that can force the selection of variants from among our own genes is a force to be reckoned with: malaria has pushed the natural equilibrium far in its favor, and to date all our efforts to push it back have failed. Even with human genetic resistance as a firebreak, the malaria parasite holds the prize as the most damaging of the microbes that use the two-species strategy for getting from one person to the next. Today about three hundred million people carry the parasite in their bodies on any given day, and untreated infections are fatal in up to twenty percent of adults and a much higher percentage of young children.

HIV, the human immunodeficiency virus that causes AIDS, has been at least as clever as tuberculosis or malaria at evading our best efforts to tame it. Its strategy of infection and contagion combines tuberculosis's capacity to hide within a cell with smallpox's and malaria's capacity to travel from person to person in a small drop of fluid. HIV needs no insect vector; any actions that carry it from one bloodstream to another will allow it to set up a new infection. The speed with which HIV spreads in a human population is determined by the most efficient of such actions: in Zimbabwe, widespread, unprotected sex with many partners has been the main route of transmission, and life expectancy overall has dropped by more than twenty years as a result of HIV mortality. In Ukraine — a new independent country experiencing all the extremes of the West in a great hurry — the percentage of intravenous drug users who are HIV-positive went from two to sixty percent in less than a year.

Once HIV is in the bloodstream, specific proteins on its outer membrane adhere to the immune system's macrophages. These immune cells readily accept the virus because, ironically, they are there to find foreign material, ingest it, digest it, and trigger the rest of the immune system to prepare for more of it. Once inside a macrophage, the virus manages to integrate its genome into the chromosomes of the cell, and from there it releases great clouds of fresh virus into the bloodstream. Other immune cells respond to the free virus as well as to the infected macrophages. The relationship between HIV and the immune system is quite ordinary at this point, as the immune system's response to the outpouring of HIV generates the fever and sweats of an early HIV infection. In time the immune system sweeps the blood clean of most of this virus, but — and here is the big but — because HIV is growing in the macrophages themselves, the immune system can never completely rid itself of HIV. Even one infected macrophage will release enough virus to infect many others, and so long as there are macrophages in an untreated HIV-positive person, there will be more HIV grown and released into the blood.

Genetic variation is very high during the reproduction of fresh HIV, and among the variants are some that can directly infect other cells of the immune system, in particular those called T-cells. The immune system is still active against HIV, and because it uses its T-cells to attack HIV-infected macrophages, there is a strong selective advantage for any viral variant that can grow directly in T-cells. This sort of variant inevitably overtakes any other versions of HIV in the body. T-cells are required for any immune response; when they start dying, the immune system crashes and the person develops full-blown AIDS.[4]

The thought of infectious microbes penetrating our bodies against our wishes can arouse deep fears of losing control over the primary boundary that separates us from the outside world. The difficulty with the antibiotic counterstrategy is that it responds to the fear of microbes invading an otherwise pristine body far more effectively than it does to the actual threat posed by microbes. The threat of microbial diseases is not that they cause a loss of bodily purity but that they upset a balance previously established between an immune system and the other microbes with which it is already engaged. So long as an early, unconscious fear of invasion impels science to concentrate on killing every last invading bug, real microbes will be able to respond as they always have, branching off into new strains that live quite well in the presence of any drug we throw at them. Antibiotic resistance is not merely a flaw in the initial strategy of total war against the body's invaders. It reveals that this strategy is based on a willful misunderstanding — a denial — of the data.

The immune system does not attempt to kill every last microbe that it detects; that would only select for increasingly dangerous survivors, as antibiotic drugs do. Instead, the immune system domesticates the microbes it sees, allowing us to live with them, and using them to keep itself primed for future invasions. A vaccine is a domesticated microbe: like a poodle or an ear of corn, it cannot grow in the wild but needs our intervention to survive. In this sense vaccines have the imprimatur of natural selection, and antibiotics do not. The inability of biomedical science to acknowledge this fact is a good example of the consequences of a science in denial.

New, virulent microbial diseases keep showing up, and we keep discovering that one or another chronic disease we thought had other causes hides an infectious microbe. Meanwhile, despite our best attempts to kill them off with drugs, at least two microbes — the agents of tuberculosis and malaria — continue to infect a considerable fraction of all people alive today. Yet the agenda for basic research on infectious diseases is still heavily invested in the dream of beating microbial enemies into total submission, as if they were small nations subject to our rules of military engagement. The dream — and the denial it engenders — keeps biomedical science from focusing on microbial domestication. The magnitude of the microbial threat to our species requires that we wake from the dream of conquest and learn how we might steer the powerful, disinterested force of natural selection toward safer, mutual accommodation.

Today's Bug Wars strategy suffers from the same flaw as the defunct Strategic Defense Initiative, the Star Wars defense partially mounted at great cost by the United States in the late 1980s. Even one Soviet missile would cause insupportable damage, and no SDI system could guarantee that every last Soviet missile would be stopped. Similarly, we can be sure that the mutability of microbes will always allow a few to slip through our chemical defenses. Going beyond Bug Wars to a mature approach to infectious disease would require redoubled efforts to develop and distribute new vaccines as well as mobilization on two other fronts: the worldwide reporting of outbreaks of infectious disease and the wide dissemination of what we already know about epidemiology and preventive medicine.

Preventing disease saves not only lives but money. In 1967 smallpox was costing us about a billion dollars a year worldwide. The thirteen-year effort to eradicate it by vaccine and education cost less than a third of a billion dollars, and every year since the economies of the world have saved the equivalent of a billion uninflated dollars by this one act of international cooperation. Despite that precedent, the United States spends less than one thousandth of its annual health care expenditures on international efforts to prevent infectious disease; that is even less than ten percent of the annual increase in overall medical expenses. Medical school education in most countries also reflects this lack of interest in public health, epidemiology, and vaccine-focused, preventive medicine. In the United States, for example, the NIH will pay all medical school tuition for a student deemed bright and motivated enough to enroll in a joint M.D.-Ph.D. program, but there is no parallel track for people qualified and motivated to get both an M.D. and a degree in public health.

The three tools of public health that can keep microbial agents at bay — an enhanced immune response, clean drinking water, and insect control — were developed to a high level of technological sophistication more than a century ago. Vaccines prepared the immune system before infection so that an incoming agent was met and domesticated before it could create a colonizing nest of offspring; clean water cut the oral-fecal path of infection for the many microbes that pass through the body; insect control kept at bay, not just the insects themselves, but any microbes that need to travel in them from one warm-blooded host to another. These tools were very successful. Yet, at the first victories of antibiotics, all three were set aside and forgotten as research strategies. As the twentieth century draws to a close and the great strategy of killing infectious agents and their vectors continues to breed its own failures, we need to recover the memory of those earlier successes.

The cost worldwide of doing nothing — or, rather, of continuing to do what we do now in the developed West — is millions of deaths each year. Five million babies, for example, died in 1992 of serious childhood diseases like measles, tetanus, diarrhea, whooping cough, and polio because they were not properly vaccinated. That is actually an improvement: in 1983 these diseases killed about nine million babies. The estimated initial cost of a serious effort to produce and distribute the necessary vaccines to protect every newborn from birth is about $25 billion, or about $5,000 per young life saved. After the first few years the cost per saved life would drop precipitously. We already know all we need to know to do this. Surely there will never be a discovery to be made from basic research that would provide so many people with such a great extension of their life expectancy.[5]

The dream of science — to blur the distinction between internal and external time and thereby to transcend individual death — is no dream for microbes. Our sort of internal time does not exist for microbes and never has; they must either divide or die.

The reward for mindless quickness is robust genetic variation: the smaller the microbe, the simpler its genome; the simpler the genome, the rougher the copying mechanism that replicates it; the rougher the mechanism, the more errors; and the more errors, the more chance for a descendant that will survive any environmental stress. While any individual microbe has its place in the sun — or the gut — for only a few minutes or hours, that very evanescence all but guarantees that among its offspring will be some — or even one — descendant different enough to live and propagate in turn, despite anything we may do to kill them all.

Biomedical science has yet to accept these fundamental differences between the microbes and us; we are still fighting a war that cannot be won. When you fight a war you cannot win, you may still aim to survive by wearing down the enemy until they are willing to accept a compromise. But today even as wise a counselor to the biomedical research community as Daniel Koshland, the editor-in-chief of *Science,* still sees the microbial threat in a uselessly anthropomorphic way:

> Humans should not confuse themselves. This is true biological warfare, in which new drugs designed by humans will become obsolete through bacterial mutations, only to be replaced by human drugs and bacterial mutations in a see-saw battle. The days of soap and boiling water to fight bacteria are long gone (although soap and boiling water are still useful) and the days of miracle drugs and universal vaccines are going. A long struggle with a premium on basic research to improve our stratagems and applied research to develop new magic bullets is clearly the prognosis for the future.

Koshland sees the problem, but he advises fighting harder, not changing the strategy. To call for more magic bullets while admitting their long-term risk is an especially sad example of scientific denial at work. There will be no magic bullets that do not selectively breed equally powerful magic microbes aimed right back at us. No matter what we do as a species, nothing in the history of life assumes that we — or any other species large enough to see — can ever reach an equilibrium with microbes in which any person's survival is assured. Because of their smaller numbers of individuals, visible species have longer, rather than shorter, odds of survival than microbes do: the Dutch elm disease fungus makes a cloud of spores and thrives; grand elm trees are reduced to shoots and stumps.

As the Nobel laureate and microbiologist Joshua Lederberg put it,

> We are complacent to trust that nature is benign; we are arrogant to assert that we have the means to except ourselves from the competition. But our principal competitors for dominion outside our own species are the microbes: the viruses, bacteria, and parasites. They remain an interminable threat to our survival. . . . At evolutionary equilibrium we would continue to share our planet with our parasites, paying some tribute but deriving some protection from them against more violent aggression.

The invisible, evanescent lives within us will continue to use their ancient weapon of mutational variety to grow inside our bodies, treating people for all their national and cultural differences as if they were interchangeable culture vessels. Their capacity to transcend our differences invites us to consider the harm we do to ourselves by forgetting that we are after all a single species. Planet-wide defenses — vaccines, public health and preventive measures — cannot be set up unless nations and peoples find a way to surrender the dream of victory that has informed our thinking about infectious disease. Short of that, creatures with no internal time at all are likely to make sure that no matter our income, language, religion, or sex, few of us will have as long a run of internal time as we might wish to enjoy our big-bodied intelligence and resources.

5

The Fear of Insurrection

> World, world, O world!
> But that thy strange mutations make us hate thee,
> Life would not yield to age.
>
> — Shakespeare, *King Lear*, IV.i.11–12

> What I'm really interested in is whether God could have made the world in a different way; that is, whether the necessity of logical simplicity leaves any freedom at all.
>
> — Albert Einstein, quoted in Penrose, *The Emperor's New Mind*

JUST AS EVERY INFECTION is an invasion of the body, every tumor is an insurrection from within. The cells of the body share a single heritage as the descendants of one fertilized egg, but the body is not a democracy. The human body is a society of a million billion cells, each living under the internal totalitarian control of its differentiated genome. No cell has free will. Most are destined to sacrifice their lives for the sake of the greater good of the body, differentiating — filling themselves with proteins particular to one tissue and no other — until they no longer can divide, then dying in order that they may be replaced by younger, fresher versions of themselves.

Some cells in every tissue sit on the sidelines, waiting for the death of a differentiated cell in order to replace it. This Greek chorus is made of each tissue's stem cells. When a stem cell divides, it produces daughters with different fates: one remains a stem cell while the other must differentiate, even though that will mean its death. Beneath the linings of the skin and gut, differentiation happens quickly. Differentiating, sterile daughters of stem cells constantly fill in the gaps left by the sloughing off of their dead, fully differentiated cousins. In other tissues, stem cells may have to wait a long time before stepping up to their task. Stem cells called fibroblasts pervade every tissue, waiting in suspended animation for the moment that they are called on by a wound to form a scar. Once stimulated by hormones released at the site of a bleeding cut, they begin to divide and differentiate, sealing the damage with a scar made of collagen, a rubbery, all-purpose protein they secrete.

The body pays a price for this ability to draw on stem cells to repair and refresh itself. A mutation in any of hundreds of different genes may allow one or both daughters of a stem cell to break free of the constraints of differentiation. Thereafter, the genetically altered descendants of that cell will continue to divide, possibly forming a clone of mutant cells, each capable of unlimited further division. When the clone gets big enough to disrupt the architecture of the tissue it displaces and to spread its descendants through the bloodstream to distant parts of the body, we recognize it as a malignant tumor.

The bodily insurrection and betrayal by a cancer are terrible to contemplate: if the body and the person are the same, how can one betray the other this way? Yet it happens, and not infrequently. About one in three of us will develop a cancer in our lifetime. Like all other failures of the body that occur primarily late in life, cancers are not of any particular interest to natural selection, so we find ourselves inhabiting bodies that are quite susceptible to betrayal as they age. Each year, about a million and a half new cases of cancer are diagnosed and more than half a million people die of the disease in the United States; worldwide, cancer kills another seven million men, women, and children each year.

A cancer cell and an infectious microbe have a surprising amount in common, even though no cancer cell ever gets beyond the body in which it was born except when it becomes the object of a scientist's passion and is kept alive in a dish. Microbes and cancer cells are both able to use the victim's body as a culture medium in which to grow indefinitely, both can stimulate an immune response, both are genetically malleable enough to provide for the chance of Darwinian selection of variants able to escape the immune response they stimulate, and in both even one escaped cell may be the source of later disease. For at least the past fifty years, the main thrust of biomedical science has been to describe the molecular differences between normal differentiating cells and their mutant, cancerous cousins and then to use that information to devise more precise ways to kill the latter while sparing the former. This response to the problem of cancer resembles to an uncanny extent the current strategy for dealing with infectious disease by searching for new families of antibiotics.

Current techniques for killing tumor cells with radiation and chemicals create the same Darwinian natural selection that takes place in the body of a person struggling with malaria or TB. As the tumor grows, throwing off a cloud of genetic variants, any mutant cells that can survive the body's defenses and medicine's assaults become the seeds of new, resistant tumors. Sometimes such mutants are overcome, and the tumor is eradicated. In other cases, the downhill slide ends with a painful death, a Darwinian catastrophe for tumor and victim alike.

These similarities between cancer and infectious disease have generated similar dreams in the collective imagination of the scientific fields dedicated to understanding their respective causes and effects. Just as the fear of bodily invasion was met by the dream of purifying the body from all microbes, a similar fear that at any time, in any place in the body, mutation in a single invisible cell may bring down the entire body has been met by the dream of being able to kill every last tumor cell. Until very recently that dream — as familiar as the dream of the perfect antimicrobial antibiotic — informed much of cancer research. It has yielded an equally familiar mixture of success and stalemate. As erythromycin once helped my body recover from pneumonia and as new drugs called protease inhibitors help my HIV-infected friends stay alive and reasonably healthy, chemotherapy, surgery, and radiation relieve many people of the tumors that would otherwise kill them. But just as some people will die of secondary infection by antibiotic-resistant variants that were selected by the very drug intended to cure them, so do many cancer victims succumb to the progeny of drug-resistant, radiation-resistant variants of their tumors.

One aspect of cancer makes it a different sort of medical problem from any infectious disease and puts the dream of total victory wholly at cross-purposes with the actual possibilities for medical intervention. Cancers arise by mutation, and most mutations can be kept from happening in the first place. As a result most cancers — unlike most infectious diseases — are avoidable. Only a few percent of new cancers are the consequence of an inherited condition, and only a few more percent are the product of infectious agents. The ones that arise from infection can be prevented as well, by curing the infection. Eliminating the bacterial cause of stomach ulcers, for instance, also eliminates the associated risk of later stomach cancer.

All remaining new cancers — nine out of ten, or more — will be neither caught nor inherited. They will be the result of avoidable habits and preventable exposures that, given the will, can be changed at any time without the need for any further basic research. Tobacco smoke is the classic avoidable inducer of cancers, but far from the only one. Foods laced with pesticides, pollutants in the air and water (both at work and at home), radiation and drugs that cause mutation: all of these cause cancer, and all can be avoided. The risks of cancer from any of them are cumulative, so cancers tend to appear in older people. Thus prevention requires the earliest possible intervention. The same mother's milk that concentrates protective immune cells and antibodies also concentrates these chemicals and delivers them to a nursing infant, where they can reach much higher concentrations than are typically found in adult tissue.

The irony is that the science of preventing cancer is simpler and easier than the science of curing it. Prevention works, and it has no clinical side effects. With very little in the way of either cash or cachet, the strategy of prevention — changes in diet, reduction in tobacco use, exercise programs — has led to a modest overall reduction in cancer deaths in the 1990s. Four percent fewer men and one percent fewer women died of cancer in 1995 than 1991. Perhaps a few lives were saved by genetic detection coupled with prophylactic surgery, but most were saved by people changing their habits to avoid cancer in the first place. Most escaped by staying away from tobacco. The different behaviors of men and women demonstrate this: a few decades ago, women took to cigarettes in great numbers as men were pulling back. In the 1990s, as the rate of lung cancer in men declined by more than six percent, it increased by almost the same percentage in women.

Every cell's DNA is vulnerable to mutation by any chemical that can bind to it and either break it or shift around some of the bonds that hold it together. Mutagens that can do this get to the tissues of our body in the food we eat, the fluids we drink, the air we breathe, and the materials we handle. Some mutagens — like those in the nitrogen compounds we breathe when the air is smoggy — are artifacts of our technology. Many others are "all natural" and oblige us to protect ourselves from them by the very sorts of chemicals that may cause further damage. One natural substance, the potent molecule aflatoxin, is made by a mold that lives on damp, stored peanuts. Aflatoxin will mutate genes in the liver cells that try to detoxify it; liver cancer can result from eating peanuts that have not been treated with the chemical pesticides that — so far — kill the mold.

The unconscious fantasy that motivates much of today's cancer research is plainly visible in any country's budget for cancer research. Prevention is hardly mentioned. Instead, genes associated with higher risk are sought, on the premise that one day the information will provide better drugs to kill every last cell of the tumor that will inevitably arise. This agenda is woefully incomplete at best and absurd at worst. For instance, to discover precisely which chemicals will cause cancer when they enter the bloodstream, and then — instead of working to remove these chemicals from everyone's food, air, and water — to study the genetics of the liver proteins that detoxify them, is to be in a waking dream.

Driven by this dream of total victory after total war, scientists have wrapped cancer research in a variety of military metaphors over the decades. Each change in the scientific agenda of cancer research has been presented as a restatement of this war aim: to kill every last cancer cell. Untethered to the reality of cancer as avoidable, the rhetoric of cancer research has changed as well, always realigning the strategies of the cancer war with the nation's priorities. We have been through three different wars on cancer of this sort and are currently in the midst of a fourth, genetic, war today.

In the 1930s, the diagnosis and treatment of the disease were left in the hands of local doctors and hospitals. The major national charity concerned with the disease, the American Society to Control Cancer, opened the first war when it began to use volunteers to raise money for a small number of "cancer hospitals" around the country. In the depths of the Depression, with armies being formed all through Europe and Asia for what would be World War II, the charity moved from the genteel to the military. Its president recruited support from the General Federation of Women's Clubs to form the Woman's Field Army, the WFA. This female organization, with two million mostly middle-class women for troops, and with officers from Commander Eleanor Roosevelt to thousands of captains and lieutenants throughout the country, was organized to fight cancer not with guns nor with scalpels but with knowledge. Learning that the disease was almost always fatal even in the best hospitals, and knowing that the few cures and remissions were almost always accomplished for cancers detected at an early stage, the WFA fought cancer with posters, radio announcements, newspaper advertisements, and the like, stressing self-examination and early detection as the best — if not the only — weapons against the disease.

The first war on cancer came to an end as World War II did, when Congress, prodded in good measure by the leaders of the WFA, created the first federal establishment for basic research on cancer and for developing new treatments, the National Cancer Institute. This new laboratory, funded by Congress with money from what had been the Office of Scientific Research and Development during the war, soon had millions of dollars for cancer research. Private funds also increased, especially after Mary Lasker — whose healthy fortune came from her family's success at creating effective advertisements for, among other items, cigarettes — disbanded the WFA and converted the American Society to Control Cancer into a fund-raising operation, the American Cancer Society. The second war on cancer was nuclear. Radiation was to be not only a fearsome new weapon of destruction but also a clean new knife that would remove all traces of the disease from a sick person's body. *Life* magazine's way of explaining this new treatment was to show a cancer cell under a mushroom cloud. As an editorial in a 1947 issue of the *Journal of the American Medical Association* blithely put it, "Medically applied atomic science has already saved more lives than were lost in the explosions of Hiroshima and Nagasaki."

This second cancer war lasted as long as the first one, all through the Korean War, the early Cold War, and the period of nuclear proliferation, until the early 1960s. The side effects of radiation had become clearer by then. The appearance of radiation-resistant tumor cells in many patients meant that this war, like the propaganda war, was not going to end in complete victory. The answer was a new declaration of war, in terms well adapted to the politics of the day: a war against invasion by aliens. This third war began about the time that the first nuclear test-ban treaties were being signed by the Cold War superpowers, a time when competition between them shifted to proxy wars fought on the territory of small Third World countries like Vietnam. In 1962 the Rockefeller University scientist Peyton Rous was profiled in *Life* in a cover story headed "New Evidence That Cancer May Be Infectious."

Even though neither the most fearsome aspect of the disease — its capacity to strike at random — nor the striking propensity of some families to develop one or another kind of tumor fit easily into the model of contagion, the metaphor fit the times. It was the centerpiece of the successful campaign, led by Mary Lasker and the American Cancer Society, to establish the twenty-one comprehensive cancer centers throughout the United States. These were backed by the heavy artillery of the third war, the Special Virus Cancer Program, a vastly expensive federal effort to find and eradicate — search and destroy — the viruses that were now argued to be the cause of much human cancer. To win the war, the budget of the National Cancer Institute was increased from $140 million to $1 billion in a decade.

By 1975, the war on cancer viruses was going no better than

the war on Viet Cong insurgents. Neither war was going to be won by the forces of the United States government, and Congress was no longer willing to support either. By 1980, the Special Virus Cancer Program had been partially disbanded. The remainder was kept going more as a support program than a high-priority research effort until a real virus, HIV, began to kill people. The AIDS crisis gave the government a clear use for the program, its facilities as well as its rhetoric. Although little in the way of curing human cancer had come out of the third war, many of the facts about how HIV works and many of the agents that slow its progress came from the redeployment of its weapons and troops. About a tenth of the NIH budget, about $1 billion per year, has gone into research on AIDS in the 1990s.

The fourth war on cancer, declared as the Cold War ended, focused on the discovery of genes that are specifically mutated in cancers. These genes liberate a line of cells to grow from a differentiating tissue cell that otherwise would have died without offspring. The genes fall into two classes: genes that would lead to the death of a cell that has suffered significant damage to its DNA, and genes that would lead to the death of a cell by normal, terminal differentiation. One gene of the first class, called p53, makes a protein that will make a cell that has received damage to its DNA commit suicide, shriveling it up in a process called apoptosis. The p53 protein is missing or inactive in a remarkably large fraction of all human tumors, regardless of tissue. In the absence of this rather severe quality-control agent, a clone of cells will not only be able to grow despite mutations suffered throughout its chromosomes, but it will also be able to accumulate new mutations — including ones conferring resistance to chemotherapy — at an escalating rate. Cigarette smoke contains many chemicals that damage DNA. One in particular — benzo-alpha-pyrene — binds best to a stretch of DNA that happens to be present in the p53 gene. As a result, many tumors in the lungs of smokers begin from — and inherit — the same mutation in their p53 gene.

In the fourth war, the enemy ceased to be an alien and became a home-grown, misguided, camouflaged militia fighter. This most recent shift in emphasis in the military metaphor is particularly inward and repressive. Today's war on cancer is heavily invested in checking IDs and passports; that is, it recognizes the genetic differences that create and identify cancer cells, so it can better monitor the success of treatments designed to root them out and kill them. An unexpected second front in the fourth war was opened in the 1990s with the discovery that some of the genetic differences that convert a normal cell into a cancer cell could also be inherited through the human germ line. Families in which such a mutation can be inherited have remarkably high frequencies of particular sorts of cancer. While avoidable stem-cell mutation is responsible for the vast majority of cancers and germ-line mutation is responsible for only a few, scientists on this new front imagined all cancers to be problems of family inheritance rather than random misadventures. Instead of being a threat to all and therefore a social problem, cancer became a personal problem for those at higher risk because of their genetic endowment.

The earlier cancer wars produced two-edged weapons, but at least drugs and radiation did stop some tumors in their tracks. Human germ-line genetics has not yet added much to the arsenal of such weapons, but it has already given new force to the old business of fortune-telling. DNA-based fortune-telling may serve as an early warning for someone who will then be able to catch a genetically inevitable clone of cancer cells as soon as it appears. But it can also mislead others into thinking that because they are free of a germ-line mutation, they are somehow protected from the more common sorts of cancer that grow from random stem-cell mutations.[1]

Avoiding cancer is a better and cheaper outcome than curing cancer. By focusing on the human genetics of susceptibility to cancer, scientists have escaped into their own fantasy of avoiding the risk of the disease. As a result, cancers that might be avoided by all of us today if we took the necessary political steps are coming to be regarded as the unavoidable future consequences of someone else's genes. Genetic medicine will be more than a fortune-teller only when the information it yields is coupled to a mature strategy for preventing the vast, preventable majority of cancers. We are not there yet, and it is difficult to see how preventive strategies can become part of a current genetic war on cancer in a country that has not yet even made medical care a right and privilege of citizenship.

Some genes are not active at birth; when they are defective, the problem is inherited as a risk of having a late-onset disease. Today, there is no need to wait for such genes to express themselves in order to know which versions one has inherited. DNA analysis can answer the question from any cell of the body at any time. With this development, genetic medicine has acquired the ability to change a person's internal time in an altogether new and unexpected way. Once you learn that the versions of a gene you have inherited will precipitate an illness later in life, the time remaining from the moment of prognosis to the moment of first symptoms will be shadowed by this knowledge. You may gain something of value by this loss when the knowledge leads you to take some set of protective actions. Without such actions, the loss of innocence is profound; in either case, it is irretrievable.

With their faulty gene an open book, families who have learned to track the occurrence of sickle-cell disease or cystic fibrosis in their offspring are now able to use DNA diagnosis to predict the clinical status of a fetus soon after fertilization. For such a family to get a positive report — that they can look forward to a child free of the symptoms that already devastated the life of a sibling or near relation — DNA diagnosis is a very good thing indeed. But for other affected families, the discovery of two faulty genes in embryonic DNA will bring a prognosis that forces a sharp, difficult decision. For the foreseeable future, such families will have only one choice — whether or not to terminate the embryo — with no hope of either treatment or cure in time to make that choice easy.

The development of the technology to read the future in a person's DNA has been so rapid and diffuse that it has some of the properties of an infection: we are now at risk of knowing our future without wanting to, without knowing why we must, and without any idea of how we will deal with the knowledge. What will our options be when we are confronted with germ-line genetic information about ourselves that would have been, in other circumstances, grounds for termination? This question, with no simple answer, is being addressed by more and more families each year as the versions of genes responsible for hundreds of inherited diseases are read and the differences converted into easy DNA diagnoses.

Human germ-line genetics and the genetic approach to cancer treatment should not have much overlap, since cancer is so common. The commonness of cancer, and the fact that all families are susceptible to it to a greater or lesser extent, tell us that cancer usually will not arise because of a recessive difference in a single gene. Yet the search for genes associated with a higher than average likelihood of developing a tumor has been vested with magically high expectations on the premise that one day it will somehow lead to better treatments.

A sure sign of the unconscious at work is the inability to give up a set of useless behaviors or habits that serve to alleviate a hidden fear or need. In this case, the hidden fear is obvious: scientists do not want to get cancer. For a scientist to be able to diagnose certain death in a number of strangers while remaining personally free of the risk and unfazed by the information, confirms the dream that science can be a way to avoid death. Dangerous variants of growth-control genes are rare except in tumors; only a small fraction of the population inherits them. The tools that detect inherited propensities pick up very rare events, so the scientists who develop them are unlikely to be given a bad prognosis by their own hand; it is only a little bit magical for them to think that they have warded off the disease. As a result, a preventable disease that can strike anyone has begun to be reinterpreted as an inherited disease, one that cannot be escaped if one was unlucky in one's ancestors.

More than one in nine women in America — and a considerably smaller but not trivial number of men — will be afflicted with a tumor of the breast; current treatments are notorious for being harsh, and the disease has a tendency to recur in survivors of available treatments. While only a few percent of breast cancers are the result of germ-line mutation, many different DNA-based tools to detect genetic predispositions to breast cancer have reached the marketplace and the clinic. These tests — which give clear results but not clear advice — detect different versions of BRCA1 and BRCA2, two growth-control genes that help to keep breast stem cells from excessive proliferation. BRCA1 and BRCA2 each exist in the human germ line in many versions, many of which are functional. Even when a nonfunctional version of either gene is inherited, the germ-line mutation has no immediate effect on the developing child. Later in life, though, children born with one functional version of either gene and one nonfunctional version are at very high risk of developing breast cancer. Only about one breast cancer in ten occurs in women — and men — carrying a germ-line mutation in either the BRCA1 or the BRCA2 gene; the other cases are due to a new mutation in a breast stem cell acquired by any mixture of choice and chance at any time.

What can a young woman do with such information that makes it worth having? Right now, and for the foreseeable future, the answer is not much. She can elect to remove the tissues at risk, but ovaries are often at risk as well as breasts, and the removal of either is a radical move with its own severe medical consequences for a young woman who may not develop the disease for another three or four decades. The knowledge gained from DNA puts these girls and women precisely in the situation of embryos whose DNA is analyzed in other contexts so that the family may consider termination: they "provide the opportunity to analyze the efficacy of medical and surgical interventions," in the words of a recent paper on BRCA2.

The majority of breast tumors begin when a single breast cell accumulates mutations in both copies of one of these genes. Telling an adolescent girl that while her sister has inherited a damaged version of BRCA1, she has not, is quite different from telling her that she will not get breast cancer. All that she can know is that her risk is the same, one in nine, as it is for other women who have two functioning versions of the BRCA1 gene. If she understands the DNA diagnosis exactly, she and her sister are in the same situation: they must be diligent in self-examination, and if they detect a lump, they must undergo an unpleasant and dangerous course of treatment, one that does not always work. If she overinterprets the result and slackens her efforts at early detection, the DNA information will actually place her at a higher risk for breast cancer than she otherwise would have been.

What have women with one functional copy of BRCA1 or BRCA2 inherited? Since it takes less time for random mutation to knock out one remaining copy of a gene than the two most people have, people inheriting the germ-line mutation usually develop their tumors at an unusually early age and have an increased lifetime risk — thirtyfold for BRCA1 and tenfold for BRCA2 — of developing a serious and sometimes lethal cancer at some time. Now that the inherited versions of BRCA1 and BRCA2 genes can be easily examined by analyzing the cells in a blood sample, all women can find out whether their risk of getting breast cancer is at the background level of ten percent over a lifetime or at the ninety percent level of those who inherit one nonfunctional copy of either BRCA1 or BRCA2.[2]

Today we give our age as the number of years that have passed since our birth. Imagine a society in which the time left for life were as clear as the time already passed. That number would go down as our current measure of age went up until it zeroed out when we did. As it becomes more refined, DNA-based diagnosis will be able to give all of us our ages as the number of years we each have left. We are already part of the way there: a forty-year-old woman with no symptoms, whose DNA has a variant version of the gene for a protein called huntingtin, knows that Huntington's chorea and death are coming in a decade or so. She is the same reverse age as an "average" person of seventy who has a statistical half-chance of ten years more of life or an adolescent with cystic fibrosis who knows that his life will be over in his twenties. The majority of people can still conclude, in ignorance of their genomes, that they are "average" and do not know whether they will make it to seventy or not. How would our society deal with the conversion of this "average" population into groups whose reverse ages are fifty, or twenty, or ten, or only one year? There is no legitimate medical purpose to obtaining the germ-line genetic information that would bring about this redefinition of age.

Whether in New York or Vermont, in fog, sun, rain, snow, or ice, each morning my wife, Amy, and I try to walk a mile up a street or road and a mile back; our exercise is at the cusp of age and circumstance, the least demanding routine we have been able to find. Our Vermont road sees a few cars and trucks and a school bus now and then as it goes past the fields and barns of our neighbors' farms. Each farm is small, with a few dozen milk cows, a stand of maples for syrup, acres of pasture hay for winter fodder, and, in the season of long days, neatly ruled stands of hard corn. Corn is the descendant of grasses that resembled the hay of my neighbors' meadows; it is a creation of human rather than natural selection, of evolution sped up by consciousness.

As soon as the unnamed ancestors of all geneticists made the brilliant discovery that the most succulent of edible grains was also able to grow into a whole new generation of ripening grass if its seeds were properly planted and watered, people were launched on the considerable enterprise of manipulating the various versions of genes in other species. Since that time, farmers have sought grasses with the biggest and most succulent heads of grain, always saving some of the seed for the next crop and eating the rest or feeding it to their livestock. Each grain was once a rarity until it was domesticated, and some could not survive at all without human intervention. Today's corn, for instance, has seeds and cobs so large and so packed that they can no longer seed themselves but must be shucked from the cob and placed in the ground in order to survive even one year.

From its beginnings, medicine has always been the enemy of natural selection. Now, with a clear idea of our genetic differences and of the risks certain DNA sequences pose, we have begun to consider a return to natural selection, but with a twist. Some scientists have suggested that we begin to directly modify specific genes of our descendants. The technology is new, but the impulse is the same as the one we put into play millennia ago: domesticating crops and cattle by selection for the traits we found desirable. As James D. Watson, who discovered the structure of DNA and was the founding director of the Human Genome Project, put it in an annual report of his laboratory:

> If we could use genetic analysis to help work out the biochemical pathways underlying memory and clear thinking, for example, we might be able to find pharmaceutical compounds to

> improve these most needed human attributes. Thus, those who want to protect the mentally ill or the slow learner may not get what they strive for if they portray them exclusively as victims of their environment. We might like to think otherwise, but only by reducing the differences in human beings will we ever have a society in which we can effectively view all individuals as truly equal.

Watson is writing about differences in mental capacity, but he may as well have been writing about any other inherited differences that put early limits on health or longevity. In medical terms, the cultivation of people can never be made into a reasonable goal, no matter what the motives. The cultivation of our own species is the opposite of medicine: the doctor's obligation is to care for all human lives, no matter what their genetic endowment. Watson's first sentence contains the way out of the trap he sets by the end of his paragraph: only by applying our knowledge of the genetic differences among us to the task of developing treatments that ameliorate the consequences of these differences can germ-line human genetics serve the purposes of a humane and fully awake medicine.

Though the differences in DNA sequence between any two people underlie and provide a biological foundation for the notion of human individuality, individuality is more than genetic. Because our brains are capable of sustaining consciousness and memory only by changing in response to the world, each of us is as different in terms of brain chemistry from all other people as our lives are from all other lives. Consciousness not only assures that each of us is doubly unique, it also allows our experiences — and therefore certain changes in our bodies and brains — to be remembered and experienced for many generations. It provides our species with a second channel — teaching and learning — for transmitting differences between individuals over times much longer than an individual lifetime.

Learned experience is the only biological channel of this sort we know of, one that does not need to encode differences in DNA in order to preserve them for many generations. Like the DNA-encoded differences in our genes, the different lessons learned in life can be transmitted perfectly or imperfectly; like mutations in genes, changes in what is learned or taught can become fixed in later generations. The second channel gives our species a unique advantage in competition with other forms of life: what may take millions of years and tens of thousands of generations to appear through DNA can be learned in a single generation.

While I was writing this book, I learned that my mother had stomach cancer. I had spent thirty years studying the molecular mechanisms of cancer in experimental animals and cells in a dish, so I knew this field well enough to be sure that its discoveries to date were of no use at all to her. My colleagues had diagnosed her disease quite definitively and even elegantly, but they had nothing to offer in the way of a cure. By the reverse way of calculating age, she had lived for eighty-three years but was less than one.

Surgery would have been a major strain on her body, and the chances of cure afterward were essentially zero. Chemicals would not have worked either, nor would radiation have had a good long-term effect. In any event, she did not want an operation that would remove her stomach; she was almost blind with macular degeneration and would have needed a keeper to feed her through an implanted tube after such an operation. She might have lived a few months longer if she had elected to have her stomach removed. But she argued, and I as her cotrustee had to agree, that she was entitled to spend her last months as fully functional as possible, and that an operation at her age — in her early eighties, and with a bad aortic valve — would likely leave her an invalid. She wished to die among close friends and without pain, and she came quite close to getting her wish.

In an impassioned defense of basic research written for *Scien-*

tific American at the height of the congressional budget debates of 1994, the renowned nuclear physicist Victor Weisskopf made the case for science as an agent of medicine's transformation, the bridge connecting an inadequate earlier medicine of empty altruistic gestures to a current medicine of treatments and cures: "Human existence depends on compassion and knowledge. Knowledge without compassion is inhuman; compassion without knowledge is ineffective."

For people who have an incurable tumor and for their families, there is no reason that compassion without knowledge cannot be effective; it is what they most need. For those with cancer and all other diseases that cannot be cured, the other half of Weisskopf's syllogism fails as well: certain kinds of scientific knowledge about cancer — the fact that one has inherited the certainty of developing an inoperable tumor, for instance — simply cannot be conveyed with compassion and may be intrinsically inhuman. In terms of our current science, the implications are plain. If it were not for the unconscious need of everyone, even scientists, to push the real meaning of an incurable disease away from consciousness, basic research would be able to contribute to that last stage of life instead of turning away from it in abject terror disguised as a lack of interest in a hopeless patient.

We will now consider how science may be able to address this last period of life, when the difference between the objective time of the outside world and the inside time left to a conscious person becomes infinitely great.

6

The Fear of Death

"Spirit," said Scrooge, with an interest he had never felt before, "tell me if Tiny Tim will live."

"I see a vacant seat," replied the ghost, "in the poor chimney-corner, and a crutch without an owner, carefully preserved. If these shadows remain unaltered by the Future, the child will die."

"No, no," said Scrooge. "Oh, no, kind Spirit! say he will be spared."

"If these shadows remain unaltered by the Future, none other of my race," returned the ghost, "will find him here. What then? If he be like to die, he had better do it, and decrease the surplus population."

Scrooge hung his head to hear his own words quoted by the Spirit, and was overcome by penitence and grief.

"Man," said the Ghost, "if man you be in heart, not adamant, forbear that wicked cant until you have discovered What the surplus is, and Where it is. Will you decide what men shall live, what men shall die? It may be, that in the sight of heaven, you are more worthless and less fit to live than millions like this poor man's child. Oh, God! to hear the Insect on the leaf pronouncing on the too much life among his hungry brothers in the dust!"

— Charles Dickens, *A Christmas Carol*

According to all scientific evidence, death is final. Resurrection and reincarnation are ideas that have no demonstrable reality, and the irreversibility of an individual death is as well established as the continued persistence of life itself. Nothing in science is proven, everything "known" has just been shown to be extremely likely — but the persistence of life and the irreversibility of death are supported by complex webs of interlocking observation with no reproducible evidence to the contrary. These facts about death notwithstanding, the deeply rooted wish not to have to die remains strong.

The earliest written monument we have is a record of just this wish. A five-thousand-year-old stele honoring the great Gilgamesh tells of his many conquests, but it also wistfully reports his failure to hold on to the one trophy that would have given him — and his subjects — immortality. The Greeks of three millennia believed that their gods might have conferred eternal life on any man or woman if they chose but that they chose instead to avoid the competition. The Greek gods did not grant immortality even to their own children by mortals; in ancient Greek genetics, mortality was the dominant allele. Of the three Fates who spun out the thread of a person's life, the one who cut it at death was named Atropos — "not changeable" — because death was understood to be inescapable.

The Greeks as well as the Romans were also quite clear that immortality would be a curse unless aging could be halted, as shown by the cautionary tales of Tithonus and the Sibyl. Tithonus was the mortal lover of Eos, the goddess of dawn. Eos asked Zeus to make her lover immortal but forgot to ask for his perpetual youth. Tithonus gradually became so shriveled that, tiring of him, Eos turned him into a cicada and shut him up in a cage. Apollo granted the Sibyl — an inspired Greek prophet — a life of as many years as there were grains in a handful of dust. In Ovid's *Metamorphosis,* the Sibyl too suffers from extreme old age, having forgotten to ask Apollo for persistent youth, and in the *Satyricon,* the Sibyl is found at last in a bottle; when children ask her what she wants, all she says is, "I want to die."

•

In 1905 Columbia University built a magnificent brick and limestone palace of science, Schermerhorn Hall, for its new and expanding departments of geology, botany, and zoology. Carved on its facade is the inscription "Speak to the Earth and it will teach you." To someone who has studied the Bible, whether the Jewish Tanakh or the Christian Old Testament, this line from the Book of Job is clearly not the motto of science that it appears to be. It is Job himself, in pain, telling his friends that neither he nor they can possibly understand the ways of Heaven and that he therefore wants to die on the spot. Appropriately enough, Schermerhorn Hall is the site of the discovery that closed for all time the chance that death could be transcended by science.

The ninth floor of Schermerhorn is now shared by the departments of biology and art history; rooms full of slides of paintings and sculptures spill into rooms full of slides of tissues and organs. In a room here in 1910, Thomas Hunt Morgan established the physical reality of a half-dozen genes, showing that a number of different genes were actually different pieces of a fly's chromosome. In this first demonstration that genes were chemicals, Morgan opened a line of research that led, in only a few decades, to our current understanding of DNA-based, chromosomal inheritance as the chemical mechanism for the inheritance of variation from generation to generation on which Darwinian natural selection depends. Speaking to the Earth after the fashion of science, these followers of Morgan have unexpectedly converged on Job's vision of the natural world. Because life is chemical in its deepest essence and random in its origins, they have shown that it need have no purpose beyond its own propagation; in studying the details of the history of life, they have found that the survival of life on this planet has always depended on, and always will depend on, the death of individual living things.

It is not merely that the death of any individual organism hardly matters. It is that individual deaths are essential: random variations in DNA that arise in one generation can enter the competition for survival only through succeeding generations. Each individual death means the loss of a singular version of DNA, to be sure, but for a species to survive, the individual members of the species must die. Science claims to control what it can understand, yet death is one aspect of life we can understand all too well without any experiments and the one over which we will never gain an iota of control. Faced with these facts, both biology and medicine have become stuck in a long series of persistent, clever, but useless attempts to ignore them: medicine, by insisting that death is its failure, and biology, by insisting that death is not interesting.[1]

The dream of immortality lives on as a way of avoiding a confrontation with personal mortality. The great fuss generated by Dolly, the cloned sheep, was, after all, driven in large measure by the presumption that people could use the same technology to achieve a kind of immortality. Setting the political question of what sort of society would lease or take a woman's body for a year for the purpose of testing the quality of a cloning technology in humans, the notion of attaining immortality through sequential cloning is misguided. It is not that Dolly is in any way her mother, any more than one twin is another. Clones, like identical twins, may begin with the same genomes and with genetically identical brains and bodies, but the moment they hit the world, they become irrevocably unique and wholly mortal people.

Setting cloning and other dreams of immortality aside, the end of life presents medical science with two large and largely untouched challenges: to diminish the effects of aging and to respond to the needs of the dying. Research agendas have little to say today about the first of them and almost nothing to say about the second. In *Setting Limits* (1987), Daniel Callahan, a wise observer of science and medicine at the Hastings Institute, proposed that medicine and science would work better if they could acknowledge that this is a world of limited resources. His last suggestion was that science and medicine both accept the reality of death:

> Not more life but a better life is an attainable goal, one that will benefit the young and the old. The third proposal is that we try to enter into a pervasive cultural agreement to alter our perception of death as an enemy to be held off at all costs to its being, instead, a condition of life to be accepted, if not for our own sake then for others. As it happens, that is hardly a new idea. It has just been forgotten or denied in practice, and it is our common task to see if it can be recovered.

A decade later, it is clear that "this common task" has not been taken up by the biomedical sciences. In contrast to the Greek myths of immortality, science may well be able to slow down aging, but it has no way to give us endless life. In their subconscious denial of this fact, many scientists and doctors have turned away from the problems of aging and dying, as if these aspects of life were somehow contaminated by what inevitably follows them. Even the last moments of life can be more than the precursor of death. What would the aging and dying look like if medical science were to see them as the last two stages of a life rather than the first two stages of a death?

The outward signs of aging are as easy to measure as the inward symptoms they produce: with the passage of time, cells wear out and die, and as they do, the tissues they assemble become less supple, more fragile, and more prone to failure. The diseases of the aged are largely if not entirely the consequence of the deaths of the stem cells that normally replace the specialized cells of our tissues. We may prod ourselves to renewed activity with the motto "Use it or lose it," but as we age, our body's stem cells seem to have their own sad response: "Use us, then lose us."

The heart shows its age as the cells that line its blood vessels seize up and die. Wounds heal more slowly, and vision becomes clouded because old stem-cell fibroblasts fail to respond when called on to divide. Infections get more severe as marrow stem cells die off and the immune response to infection becomes slower, briefer, and less vigorous. The mind falls into dementia as nerve cells in various portions of the brain shut down and die. The stem-cell deaths that bring on aging, however unpleasant in their consequences, do not necessarily lead to death, certainly not right away, so a science willing to keep clear the distinction between aging and mortality may well be able to ameliorate the symptoms of aging.

It is not clear why a tissue's stem cells start to die after four or five decades of finely tuned division and differentiation. According to one theory, stem cells die when the DNA in their chromosomes accumulates too large a burden of random errors, and as errors are most likely to occur during the copying of DNA, the fraction of stem cells dying will go up with every stem-cell division, that is, with advancing age. An opposing model proposes that stem-cell loss in aging is no accident at all but a genetically programmed stage in the development of the body, no different in mechanism from the development of the embryo from a fertilized egg. Certainly cell death is part of normal embryonic development, and throughout life, programmed cell death remains central to the proper function of many tissues. A fetal hand looks like a mitten before it looks like a glove; as their bones form, fingers are carved out by the death of cells between them. The genetic model simply extends the notion of differentiation to aging.

From these antithetical models, a third falls out with a Hegelian thud: natural selection sets a life span, but by random error, not by specific gene action. Proofreading — the correction of errors in newly copied DNA by molecules in the nucleus of a cell — is tuned by natural selection to miss a sufficient number of errors to assure that stem cells die off so that an individual dies when the individual's survival is less beneficial to the species than its death would be. This synthesis of the two models explains the otherwise peculiar fact that natural selection has given our cells a copying apparatus that makes a surprising number of mistakes at every cell division and a separate, complicated set of proteins that work at every division to fix all those mistakes.

In the third model, the need for errors in the germ line creates individual mortality as an inadvertent side effect. This peculiarity — if it is so — would be a particularly poignant example of the difference between the survival of a species and the survival of any individual within the species. A level of uncorrected errors in the germ line is needed to keep the species from dying off through insufficient variation, yet the accumulation of genetic errors in the stem cells of the body eventually kills off the individuals in the species. This model nicely explains how two species — like bats and mice — can have the same size and overall metabolism but differ in life expectancy by a factor of three- to fivefold: one simply has better repair machinery.

A model that explains aging as the consequence of stem-cell DNA errors that were missed by a repair mechanism puts cancer in a new light and explains what would otherwise be a bizarre coincidence: at least four different inherited conditions that lead to childhood cancer and premature aging involve a mutational loss of some part of the DNA repair machinery. These conditions speed up the clock of aging that measures time by the accumulation of DNA errors: in the absence of repair, a decade's worth of damage in an ordinary cell can accumulate in a year. In all four conditions, tumors arise as the result of unrepaired errors to genes necessary for growth control. Eventually the cells of these tumors will also die, of other, rapidly accumulating errors, but that is of small consolation to the teenager who looks fifty and has a cancer of old age.[2]

These syndromes suggest the possibility that cancer and aging may well be different outcomes of the same fundamental process: the accumulation over time of unrepaired genetic damage to different, specific subsets of genes. A tumor is, after all, a collection of descendants of a cell that had an unrepaired error in one or more of its growth-controlling genes and thus had escaped the body's controls. Cancer research has given us many tools to accelerate the death of mutated stem cells. To the extent that aging and cancer have a common underlying mechanism, standing the insights of cancer research on their head may turn out to be a useful way to discover tools for delaying the effects of aging.

As with the genetic component of susceptibility to cancer, so too with aging: if this model for aging is correct, then we can expect people who inherit different versions of the genes that carry out DNA repair to have different life expectancies. And as with the environmental, avoidable aspects of cancer, so too with aging as well: no matter which versions of a DNA repair mechanism one inherits, lowering the lifelong dose of mutagens is likely, according to this model, to extend life as well as lower the risk of cancer.

The straightforward experimental work necessary to distinguish among models of aging that see it as an accumulation of random events, as the outcome of a genetic program, or as a synthesis of the two has not yet been done, and the evidence that is available is not sufficient to exclude any of our models. For example, although DNA taken from the cells of an older person's tissues will show more accumulated random damage than DNA taken from the cells of a younger person, it is not possible to say whether that damage is the cause or the consequence of the aged tissue's loss of function. Only cells that have not yet died are around to provide DNA for the test, so they are not likely to be the ones that have suffered the most serious DNA damage.

Some specific chemical changes that occur in older people's

tissues are clearly the result of programmed gene regulation, not random DNA damage. For example, from the earliest embryonic stages on throughout life, the steroid hormones estrogen and testosterone turn genes on and off in almost every tissue. Women's cells get their estrogen through the blood from other tissues; the cells of men actually synthesize it from blood testosterone. The levels of both hormones are programmed to fall in the fifties, middle age today but well over the typical life expectancy for most people until about a hundred years ago.

The decline in estrogen becomes particularly troublesome to those cells whose genomes need it to carry out their designated differentiated tasks. The neurons of the forebrain, for instance, need estrogen to make acetylcholine, the chemical they use to communicate with one another. Acetylcholine is synthesized in neurons by a protein called choline acetyl transferase, or CAT. The gene for CAT cannot be turned on without estrogen. Falling estrogen contributes to the aging-associated decline in brain function: unable to make acetylcholine, a forebrain neuron will not merely fall silent, it will also die. Why not simply give all old folks estrogen, then, to retard the loss of cortical neuronal function so sadly typical of old age? Estrogen's centrality makes it too blunt an instrument for that: other tissues would respond in unwanted ways, and the result would be a collection of serious medical problems.

If the cortex of the brain, with its store of neurons laid down for good so early in life, has any stem cells capable of producing neurons to replace the ones that die in stroke or dementia, we do not know how to find them. As a result, the neural substrates of consciousness and memory are particularly vulnerable to cell loss. In Alzheimer's disease, for example, portions of the brain responsible for memory lose function prematurely as their neurons collapse under the burden of accumulated, damaged proteins. Some cases of Alzheimer's are familial, a sign that susceptibility is the result of inheriting a variant version of one or more genes. These cases often have a much younger age of onset than the more common sporadic version of Alzheimer's, which usually sets in among people who have lived for at least six or seven decades.[3]

To get a clearer understanding of the mechanisms of cellular aging, it would be useful to be able to manipulate the cells of a tissue without disrupting the signals that keep each stem-cell division tuned to produce one differentiating daughter and one daughter that remains a stem cell. That has so far been very difficult. Unable yet to carry the architecture of a real tissue into a dish, scientists must instead disperse a tissue into separate cells. The fibroblasts that are released from the tissue in this way can adhere well to the surface of a dish; they grow into colonies as if they were autonomous, single-celled microbes of a sort. A dish of growing fibroblasts is not much more than a scar, but cultured fibroblasts have provided some intriguing data support the notion that cellular aging is itself a differentiation encoded in the genes of every tissue cell.

In a laboratory culture dish, fibroblasts from a young individual will quite suddenly stop dividing after about fifty divisions, whereas those from an older person will stop much sooner. The cells do not die at first but rather swell up, becoming the senescent, sterile last descendants of their lines. The causal relation of cellular senescence in a dish to the aging of the body is unclear, but should stem cells in the body turn out to undergo cellular senescence after about fifty divisions, that would certainly set an upper limit on the lifetime of a body assembled and maintained by the constant replacement of million of millions of differentiated cells. For example, the programmed death of immune system stem cells after fifty divisions would explain the degraded immune response of an aging person, but these stem cells are so fastidious outside the body that we cannot yet grow them well enough in a dish to find out.[4]

A new and unexpected mechanism for programmed cellular aging has recently and unexpectedly turned up in the hands of scientists studying one of the quirks of the chromosome-copying machinery. Any cell has difficulty dealing with free ends of DNA. Each chromosome of the cell contains a single, extremely long, skinny molecule of DNA. Each end of each DNA molecule is a potential problem for the cell, for two different reasons. First, the machinery cannot even start copying a straight line of DNA without first clipping a short stretch from one of the strands, leading to an inexorable shortening of each chromosome with every cell division; and second, each end could easily be mistaken for a piece of broken DNA by repair enzymes. If that were to happen, it might be connected to any other DNA end, sticking one chromosome to another and making cell division messy and eventually impossible.

Scientists have recently found the cell's elegant single solution to both potential problems: every chromosome tip ends in a disposable stretch of DNA called a telomere. Telomeres are runs of thousands of copies of a simple repeat of a few base pairs; unlike genes, they do not encode information for the production of proteins but rather serve as knobs to protect the ends of the chromosomes. Though each telomere gets chopped back with every round of DNA copying, there are enough extra copies of the sequence for that not to matter at first; proteins recognize and bind to the repeat, protecting even shortened tips from nosy DNA repair enzymes.

Because telomeres get shortened in the process of DNA copying, they serve as a clock, measuring the number of generations since a stem-cell line was born in an embryo. Once the telomeres of its stem cells are chewed down to raw ends of DNA, a tissue begins to show its age. Just as a Metrocard has one toll knocked off its value each time it is used to take a New York City bus or subway ride until it is reduced to a valueless slip of plastic, once the telomeres are gone, that is the end of that stem-cell line. It may live a while longer in the body — you can ride the subway for a long time, even on one fare — but it will never divide again.

One set of stem cells in the body — the germ line — must somehow be kept from the telomere clock's measure of internal cellular time. Germ-line cells cannot allow their telomeres to get shorter, since the entire species depends on their surviving to produce further generations. Natural selection has found a familiar solution to this conundrum. Just as the problem of fidelity in DNA copying is solved most successfully by allowing mistakes to happen and then fixing them, the problem of the germ-line telomere is solved by allowing the germ-line chromosomes to be subject to telomere shortening, then giving them access to a special telomere repair enzyme called telomerase.

At each generation, a germ-line chromosome's tips are cut back, then the tips are restored by telomerase. Telomerase keeps the sunset of cellular senescence from reaching the germ line, preserving its chromosomes in pristine condition for the long haul. Each time the telomere is replaced, the germ-line cell experiences a generation-long pause in time, quite complementary to the speeding up of time experienced by a tissue cell from a person with Bloom's syndrome. No matter how many cell divisions in the body have separated a germ-line cell from its ancestral fertilized egg cell, its telomerase keeps it ready for more divisions, more fertilizations, and more generations.

Telomeres are long enough to allow around a thousand divisions before they are chewed back to the point of cellular lethality: the stem cells that provide the short-lived cells of the body's linings and blood divide every week or so, accumulating a thousand divisions after a few decades, at just around the time that natural selection loses interest in a body. If we could find a way to turn on telomerase selectively in the stem cells of tissues and not their differentiated daughters, we might be able to delay the senescence of stem cells and thereby extend the useful lifetime of the body. Exquisite selectivity would be needed to be sure that telomerase was activated only in the stem-cell daughters of a stem-cell division and in the differentiated daughters whose fate would normally be to die.

Both germ-line and tumor cells can divide without limit, but differentiated cells cannot. The risk of cancer is the most serious possible side effect of intentional telomerase activation in a differentiated cell: spontaneous cancers may turn out to be no more than natural but particularly unwanted cellular solutions to the problem of aging. The similarity is more than coincidental: many tumor cells produce the same telomerase that is normally made only in the germ line. Once armed with illegitimate telomerase, a line of cancer cells loses its capacity to tell how many cellular generations have passed, so the cells in the cancer may grow indefinitely, never senescing. It is not clear whether telomerase is necessary for a cancer to overtake a normal tissue, but until we know that it is not, telomerase is certainly not an enzyme that one would want to see intentionally or accidentally activated in any normal cell outside the germ line.

A more complete agenda for the science of aging would include the search for ways to enhance the activity of the DNA repair machinery of a stem cell, to activate a stem cell's telomerase without causing its differentiated daughters to become cancers, and — most important, because it is least difficult — to reduce the dose of any agents that cause random errors during DNA copying. We do not need to learn anything more about the encoded, inherited aspects of aging to act on the knowledge that regardless of whether mutant repair enzymes, telomere loss, or wholly random DNA errors eventually bring stem cells to their knees, everyone's life expectancy would be extended by reducing external sources of DNA damage to the absolute barest minimum.

This is not a new idea: in *What Is Life?* Schrödinger also saw that because X-rays cause mutations in fruit flies, the gene was likely to be a large, fragile molecule. Nuclear power plant waste materials, diagnostic X-ray machines, the radon gas that leaks from granite in the ground, and cosmic rays that strike the atmosphere can all damage DNA. Fifty years later, it remains a good idea for medicine to learn how to protect a person from such radiation — and from all other sources of DNA damage. This is not to say that slowing the aging process by reducing the risk of DNA damage is simply a matter of cleaning up the environment. There will never be a way to reduce the risk of DNA damage to zero any more than there can be a way to reduce the frequency of errors in copying to zero. We inevitably make mistakes in our DNA simply by living.

Molecules that break DNA are constantly being made in each cell as by-products of the conversion of sugars into energy. Every time we eat, some portion of the food is converted to sugars that burn inside our cells. Burning breaks the bonds that connect the carbon atoms of a sugar to one another, then reassembles each carbon atom, in a new configuration with two oxygen atoms, into carbon dioxide. Because it takes less energy to assemble this new configuration than photosynthesis took to assemble the sugar from water, carbon dioxide, and sunlight, energy is left over.

In an ordinary fire, that energy would be dissipated in the random motion of the molecules involved, and we would perceive it as light and heat. Inside our cells, some of the energy leaks away as well — that is how we can maintain a body temperature above the temperature of the world we live in — but not all of it is lost. Some of the energy of burning is retained by the cell in small molecules called ATP. The energy stored in ATP is then spent, in an orderly way, to carry out the tasks of living: copying genes, making the proteins of differentiation, and sending and receiving signals from the outside world. ATP and heat energy are the safe products of the cell's burning of sugar; mutagens are its unavoidable by-product.

As the bonds between carbon and oxygen shift, highly unstable compounds called free radicals fall out of the chemical furnace and into the cell. Unless sopped up quickly by a scavenger — a molecule whose bonds may be broken with no ill effects on the cell — a free radical can easily break a bond in the DNA of a chromosome. The risk of self-imposed mutation of this sort is not as high for chromosomes in the nucleus as it is for the small chromosomes of the mitochondria in each cell. Mitochondria contain the cell's sugar furnaces, and it is their DNA that is most likely to be damaged by free radicals.

The risk of mitochondrial suicide is highest in the tissues of the body that burn the brightest, the brain and muscles. The weakness and dementia common to very old people result in large part from the damage accumulated over time to the DNA of mitochondria in their nerve and muscle cells. Inherited errors in mitochondrial DNA reduce the efficiency with which their mitochondria extract energy from food. Here, as in the case of inherited progeria with early cancer, germ-line experiments of nature point to the eventual fate of all of us: the disabilities suffered by young people who inherit damaged mitochondrial DNA usually include neural and muscular disorders.

Mitochondrial free radicals are the inevitable product of our need for energy, the overhead in aging our bodies must pay for the calories burned over a lifetime. All the cells of the body must risk this damage in the work they do to stay alive. To outwit natural selection's poisoned gift of free radicals and escape the damage they do to our mitochondria and chromosomes, we can turn the tables and eat scavengers. Vitamin C and vitamin E are both good scavengers, and green and yellow vegetables contain lots of others. We can also reduce free-radical pollution in our cells simply by eating less food. Animals kept on diets that provide adequate protein and vitamins while radically restricting calories live about 20 to 25 percent longer than the average lifetime of their species. Their brain and muscle cells have low levels of free radicals, and their mitochondrial DNA accumulates less damage than the mitochondrial DNA of old animals on regular diets.

The fact that calorie deprivation slows aging tells us that the DNA damage incurred by an ordinary diet can damage cells whose telomeres are in fine shape, and that many people age and die even though the telomeres of their chromosomes still have many cycles of copying left. People who have lived on reduced calories but healthy diets for months or years — the volunteers who lived in the controlled environment of Biosphere II in the early 1990s, for instance — look starved but feel healthy. Turning down the flame of calorie consumption has its hidden costs as well, though: eating is fun as well as necessary, and not all of us can emulate Kafka's Hunger Artist. A longer life of self-imposed, permanent hunger may be a mixed blessing at best to people who wish to eat their cake and have the time to eat some more of it.

Slower aging and longer life are both much to be desired, but they are not exactly the same thing, and it may turn out that we will have to choose which we want before we can have either. A future medical science able to delay aging would have at its disposal a set of facts that we do not yet have. At the least, it would be able to target the body's many stem cells selectively, giving them the special repair mechanisms that operate naturally only in the germ line but without perturbing the programmed cell death of their differentiating daughters, and it would know how to protect mitochondria from self-inflicted damage as they make ATP from the burning of sugar. We can get some idea how life would be for all of us if this knowledge were available by looking at the figures for people who get through their seventies and early eighties today without succumbing to any of the common degenerative diseases brought on by the death of brain, muscle, or blood vessel cells. They seem to be able to live another decade or so in relatively good health, with no more likelihood of a medical problem in their nineties than in their sixties, but then they die before they reach a hundred.

Living to this maximum human life expectancy of about a century with fewer effects of aging would be wonderful, but extending that life expectancy beyond a century might or might not be so wonderful. It is important to remember that these are different targets. Those concerned about overpopulation consider any attempt to extend our maximum life expectancy irresponsible. Our species will reach a population of approximately 10 billion sometime in the next century, and any significant increase in the maximum age would diminish the chances of stabilizing it at that level. A jump in our numbers and the average age of the population would present a new set of medical and social problems. If a pathway to an extended lifetime exists at all, it is questionable whether it would be desirable to find it. There is good reason to doubt that it exists. Natural selection has provided the members of every other species of mammal with a maximum life expectancy. With a tiny number of exceptions aside, ours has been about a century for all of recorded history; so far, medical science has not made a dent in that looming wall. It seems prudent for science to concentrate on learning how to slow the aging process rather than on how to extend the life expectancy of our bodies.

We may know how to counter the effects of aging within a few years or a few decades or never. People dying in very old age after having been successfully kept from the slow decline of aging will still want — as we do even now — any assurance they can get that the quality of their remaining lives, however short, will be preserved until the very moment of death. How many scientists and doctors would consider that last task to be of any scientific interest and worth the work? How many would be able to get past giving false promises that every condition is curable in principle, that impending death is the failure of a cure, and instead concentrate on the medical and scientific aspects of the very last stage of life — that is, on dying?

Today, with the majority of people dying before eighty of infectious, environmental, behavioral, and inherited problems we cannot yet solve, this issue may seem premature. But there is a risk involved in not confronting it now: silence means we will see biomedical research continue to use each successful reduction in premature death as an excuse to avoid dealing with death's inevitability. The old people who today suffer avoidable disease, unnecessary isolation, and pain in their dying are the major victims of this avoidance of the link between aging and death. They deserve a better deal than the one they are getting from today's medical science. Not for them alone, but also for today's children who will be the aged of 2050 and beyond, science and medicine have to learn how to attend to the problems of the dying.

The medical treatment of the dying is almost invisible today, an embarrassing situation that can only get worse as the rest of medical science succeeds in allowing a greater fraction of the population to live into old age with sufficient residual mental and physical capacity to understand their situation. For the sake of these lucky people — may we all be among them — medical science is obligated now to begin a research effort focused on making dying itself as brief, and as healthy, as possible. This is no joke: the hospice movement — not a product of scientific medicine but a reaction to it — has shown that a dying accompanied by a minimum of pain and a maximum of social interaction is healthier and better by far than the typical dying of today, accompanied as it so often is by prolonged agony and isolation.

For most of my life, and for all of my thirty years as an experimental scientist, I scrupulously avoided my own personal and professional responsibility to attend to the dying. It is not that I had no chances to make the connection between science and dying; I simply chose not to take them. In my own confusion, I lost sight of the fundamental truth that dying is as distant from death as any other stage in life is. In fact, under appropriately absurd circumstances, I attended a dead body before I could bring myself to attend to a dying person. In my first year as dean of Columbia College, I was called to a dormitory to deal with a bullet-riddled corpse that four students had inadvertently brought into their room in a rug they had taken from a Dumpster outside their building. At that time, a corpse's eerie calm was less disturbing to me than merely contemplating a visit to someone who was dying.

The dying of three of my four grandparents is wrapped in the mystery and secrecy of my childhood. As a scientist, I was able to help my paternal grandfather to live a decade longer than he thought he would; he was a pleasure to talk to, but I avoided being with him in his last days. I first got involved with a dying person when I arranged for a colleague to be removed from a major research hospital, which had given up on him, to a hospice, which was able to ease his suffering. When another colleague was dying of the drug that was meant to cure him of a particularly bad cancer, I was unable to make any useful emotional contact, and I withdrew from his deathbed filled with an impotent mixture of sadness and confusion.

The deaths of my parents bracketed the period in which I came to see how a failure to acknowledge death properly distorts the practice of medical science. My father died of a respiratory infection acquired in the hospital a decade after he had lost his senses to Alzheimer's disease. During his last years I did not see him at all, and I did not understand that he was dying, for I already imagined him as dead. He lived for many years in a home for the demented, his body kept alive by strangers because his family — myself included — could not carry the burden of caring for him after he ceased to know who we — or anyone else — were. He was allowed to die at last, of pneumonia, because my parents had signed papers in advance, asking that their lives not be extended by heroic measures once they had crossed an irreversible threshold of pain or dementia.

My mother survived him, and in her last months, and even in her last days, she gave me and my family ample evidence of the difference between dying and being dead. She became stronger as she became weaker, became increasingly generous and wise with me and my relatives, and with a host of new and old friends, in ways that she could not while she was more fully alive. This stunning emergence of a kinder and wiser person from the dying body of my mother came to a halt only in her last few days, when the pain of her tumor began to require such high doses of morphine that she was unable to speak with any lucidity. Even then, she clearly accepted her death, said good-bye, and, with the help of hospice care at home, died peacefully.

Hospice care is still controversial at many major medical centers today, for its goal is not to provide good treatment for the dying but to provide a good death. At their best, hospices excel at delivering what they promise: control over pain, dignity to the end, and the assurance that no one need spend their last moments alone. Bruce Jennings of the Hastings Institute gave this description of the good that hospice care can do; try to imagine anyone saying this of today's hospital-based, invasive interventions:

> The wisdom of hospice reminds us that all individuals will die, and that until then we all live and flourish through relationships of mutual giving and interdependency. To live well, or

more precisely to live the human good; to be healed, and to be sustained meaningfully whole — these things are to grow, to change, to be transformed. These are the goods of hospice because they are as much a part of the human life story in its final pages as they are at the beginning of the book.

The current institutional response of science to the dying reflects my own attitudes during those decades I worked in my lab. It goes something like this: "You have had the misfortune to be born too soon to benefit from science's ever deeper comprehension of nature. That is too bad, but since we can know how everything works, certainly one day we will know how to keep a death like yours from happening. Until then, you will understand if we do not spend much time on the relatively uninteresting matter of how it is to die."

Today, medical scientists treat very old age, dying, and death with equally fastidious disdain, as if they were all somehow intrinsically uninteresting. If they are as frightened of death as everyone else, then their disdain for aging, death, and dying is a prophecy that keeps them from confronting their fears. A good deal of interesting science lies waiting to be done by scientists able to admit their fears of death and look beyond them to study dying on its own terms.

The questions to be asked are familiar: which parts are painful and may therefore be made better by the easing of pain; which parts are inherited through the genome and may therefore be made better by the manipulation of the genome or the addition or subtraction of a gene or a protein; which parts are conscious, and which are unconscious, so that we may better understand how it feels to be dying and learn how to alleviate the worst of those feelings. Those questions would form a minimal agenda for research on the dying stage of life.

Beginning with Elisabeth Kübler-Ross's 1970 classic, *On Death and Dying*, many serious studies of dying have been built around interviews with people in the last days of their lives. A doctor herself, Kübler-Ross broke many rules at her hospital by insisting that the dying be given a chance to describe their feelings directly; simply allowing the dying a voice was a major accomplishment. From their narratives, she produced an anatomy of the physical and emotional stages of dying: denial, anger, bargaining, depression, and acceptance. As she points out, all but the last of these five stages express a deeper and more fundamental denial, attitudes that allow one nevertheless to have some hope. Hope in the face of certain death may seem absurd, and perhaps it is, but nevertheless the dying showed her — and many studies since have confirmed — that a dying person does not lose hope until just before death.

A person's last days can be the most remarkable example of dying as an aspect of living: without hope, a dying person begins to pull away from the world, sleeping a lot, not seeing anyone, not interested in anyone. At best, and without pain, the end of life seems quite remarkably like the beginning, the clock of internal time run backward one last time, to the earliest days of infancy. Kübler-Ross counseled that hope should never be denied, that the dying should not be burdened with facts that would remove all hope before the person was ready to set it aside, and that the enemy of the dying is not unavoidable death so much as avoidable physical and mental pain. In the decades since Kübler-Ross's book came out, about a third of her readers have passed through her five stages and died. In all that time, precious little has been added to, or taken from, her five-stage formulation of dying, and almost nothing has been done in science to carry out any of her prescriptions.

There is, then, a realistic scientific agenda for the period from the moment when there is nothing that medical science can do to stop death from coming until the moment of death. It is to understand the mind and the body well enough to keep both as free of pain, and as free of isolation, as possible. Science can complement the work of a hospice by providing it with new tools to accomplish these ends. After my mother died, I found this agenda clearly laid out in an early text, one that comes from a nonscientific tradition, quite different from the Greek tradition at the root of my science. Almost two thousand years ago, the interpreters of the Law for the Jewish people began a process of debate and deliberation that went on for centuries. The written records of these arguments are preserved in the Talmud, and in one of them I found a short discourse on the rights and needs of a *gosess* — a man who is beyond medical help and expected to die in fewer than three days. The interpreters rule that such a man may marry, may sign legal documents, and may even incur debt. In other words, healthy people are obligated to persist in their normal interactions with the dying; to do otherwise would be an act of heretical presumptuousness, an arrogation of God's singular power over life and death.[5]

Taking these ancient rights of a dying person as a guide, much dying today happens poorly, with unnecessary pain. It is time for medicine to acknowledge what torturers have always known: pain is a pathological state that mocks any pretense to health. To uncover the underlying mechanisms of pain, it is useful first to recall that no matter what part of the body is in pain, the hurt is, of course, in the head. Pain is a brain state, and as such it ought to be as understandable, and treatable, as other unwanted brain states are turning out to be. The most effective painkillers we use today work only by dulling the senses, and all are highly addictive when taken by people whose lives are not almost at an end. Doctors who try to prescribe large enough doses of these compounds — morphine and its derivatives — are often suspected of inducing a dying patient's addictive craving. This is a cruel joke to anyone who is dying with intractable pain and who may reasonably argue that one cannot be addicted when one is dead. A civilized medicine that fully accepted the reality of death would also recognize that the pain itself is as damaging as any addictive state. There is another, equally ironic barrier to the straightforward study of the proper pharmacology for intractable pain: the fear that an overdose of morphine might be used intentionally to shorten the life of a dying person, with or without the person's consent. It is ironic because the most frequent reason for requesting an early death is precisely unbearable pain.

Beyond the tragedy of dying people having to hasten their death with the same compounds that might have given them a reason to live longer, the denial of proper painkillers damages a person's body. A person in pain suffers from a reduction in the efficiency of the immune system and usually cannot actively participate in any other courses of treatment. We need a major effort to find or synthesize — and then to distribute openly — a new generation of more effective painkillers. Such research would need strong government support, since the political problems of such research and development make these studies as uneconomical as vaccine production for today's pharmaceutical firms.

People need the touch of other people's hands — those soft touches that let them know they are not alone — all their lives, to the very end. The cruelest of the paradoxical consequences of the denial of death in modern medicine is the insistence on treating a dying person in ways that destroy all chance of privacy and dignity, that deny the person the ancient right to the continued presence of friends and family. The usual argument for leaving the dying person alone in a cold room with tubes and monitors blocking all human interaction, for allowing the rarest and sometimes the richest of words to go unheard or unsaid, is that this regimen is necessary to extend the person's life, albeit only for the shortest of times. But to extend external time by so little while removing all chance of the person's sharing any of the little internal time left with anyone else is surely another form of de facto torture, equal to withholding painkillers.

If the dying who have stayed conscious were given painkillers and allowed social contact, then they would have all that we could offer them as they lived out their last moments. But what of the dying who have already lost all sense of others, all ways to measure the passage of time within themselves? A person like my father did not have the chance to use the rights granted to him by the Talmud. The "how" of Alzheimer's disease — the mechanisms of gene expression, protein synthesis, and cellular communication that work so well for a century in some brains but not well at all in others — is at the intersection of basic biomedical science and the right of a dying person to full membership in society until the last moment of life.

Daniel Callahan has made three proposals for setting limits on our medicine and science, to align them with the fact of mortality. I opened this chapter with one of these proposals — that we all accept the inevitability of individual death — and the other two bring us to the end of this book. His second proposal is that we support research that would increase the quality, not the length, of life. He then proposes that we shift, politically and socially, from a commitment to do whatever it takes to keep an old person alive to a medicine that acknowledges the finiteness of life and the certainty of death.

He does not say specifically what science or medicine might do to carry out these three proposals. There is a great task ahead: moving the curve of life expectancy of our people upward until everyone lives out a full life in reasonable health. Saving the lives of millions of young people and giving millions more decades of additional life are realistic and more humane targets than the impossible, death-defying promises of today's medicine. The goal of medical science should be not eternal life but a pain-free, socially engaged life for a full genomically allotted span of years.

Conclusion

> The despisers of mankind — apart from the mere fools and mimics, of that creed — are of two parts. They who believe their merit neglected and unappreciated, make up one class; they who receive adulation and flattery, knowing their own worthlessness, compose the other. Be sure that the cold-hearted misanthropes are ever of this last order.
>
> — Charles Dickens, *Barnaby Rudge*

> The rationale of my theoretical program is no different from the familiar clinical one of interpreting the infantile unconscious, so-called: in both cases one's objective is to facilitate acts of discovery and revision through explicit confrontation, consternation, and reconsideration.
>
> — Roy Schafer, *A New Language for Psychoanalysis*

THE GREAT EXPERIMENT in coercive social engineering that became the Soviet empire began with a striking slogan, now as forgotten as the red flag of the Soviet Union: "From each according to his abilities, to each according to his needs." The experiment was a failure, but the slogan contains an objective worth our attention. A future in which each of us got what we needed and gave what we were able would surely be a place where those whose skills and training led them to the medical sciences would be giving the rest of us a more equitable and realistic version of medicine than the one we live with today. But just as surely we will never reach that future, nor any other more desirable one, until everyone — scientists, doctors, and the rest of us — understand the particular needs of scientists and doctors themselves and decide to try to meet their special needs in turn.

Everyone alive needs to make some sense of life, to give it some meaning. The doctors and scientists who created today's medicine and who will create tomorrow's share this need. In that, they are no different from anyone else. There is a difference, though, in what they know, and that difference makes the task of giving meaning to life much more difficult for them than it might be for anyone less aware of certain facts of life.

In the past century scientists and doctors have made four interlocking discoveries that have made the task of finding meaning much more difficult. The oldest discovery has had the deepest impact: it is that DNA-based natural selection generates life in all its diversity and orderliness — including a scientist with a brain of great capacity to understand life's structures and functions — while, by itself, natural selection contains no element of design nor purpose.

The second discovery concerns the mind. Scientists have shown that the conscious mind is the product of cells in the brain, an expression of the capacity of genes in these brain cells to respond to the outside world as well as to selectively recalled memories of earlier interactions with it. Third, they have found that the brain that does this is a tissue made of cells like any other tissue, albeit one that can imagine it has — or is — an ineffable, nonmaterial soul. And most painful of all, they have found that the entropic tendency of large and complicated structures to degrade into smaller ones assures that death — including the death of the inner voice we each hear when there is no one else in the room — is irreversible.

Together these discoveries paint a coherent and clear picture of the living world and of our place in it that is notable for its complete lack of meaning. Everyone who learns of these discoveries has the double task of finding a way to accept them, despite their cumulative power to exclude design and purpose from the living world, and of helping to assure that the science of the future will be made by men and women who have found meaning in their lives despite these facts of nature.

It will not be easy. Scientists cannot simply avoid these discoveries, as so many of the rest of us do. Many aspects of today's medicine are based on precisely these discoveries, which is why medicine has come to reject any larger meaning or purpose to life beyond the workings of genes and the capricious choices of natural selection. Yet one must — or at least I think one must — see life as more meaningful than that if one is to lead a life worth living. The alternative — an unconscious rejection of one or all of these facts of life, along with grandiose promises to find a way to conquer death itself — has not worked. New data keep emerging from laboratories to make such denials and promises ever more hollow.

For example, the DNA of people suffering from Huntington's disease reveals a quirk in the human germ line that predicts an ironic end for the experiment of nature we call consciousness. There is a good chance that our species will become demented and die off from the consequence of a peculiarity of the human germ line that causes diseases like late-onset dementia. One of these, Huntington's disease, is brought on by the inheritance of one variant gene encoding a protein called huntingtin, which plays a still undetermined role in the workings of a set of cells at the center of the brain. Even though they have also inherited a fully functional huntingtin gene from their other parent, people inheriting the variant of huntingtin begin to lose mental functions in their middle years, followed by a loss of muscle control, which leads inexorably to an early death.

Huntington's disease is not brought on by a classic mutation. Like dozens of other inherited diseases affecting the brain, it begins in a parent as a stutter in the DNA-copying machinery of the germ line. Certain runs of bases in DNA — in particular, repeats of triplets like CGA — are prone to this mistake. At every generation of DNA copying, they have a chance to be accidentally reduplicated so that what is a stretch of . . . CGACGACCGA . . . in one generation can become . . . CGACGACCGACGACGACCGA . . . in the next and . . . CGACGACCGACGACGACCGACGACGACCGACGACGACCGA . . . in the generation after that. Even one reduplication of this sort is problematic for the meaning of a gene, but a string of them can turn a gene into a wholly useless mess, causing the chromosome itself to fall apart. People inheriting a moderately lengthened stretch of CGAs in one of their genes will usually show only mild symptoms of whatever problems are caused by reduplication. However, they are at high risk themselves of making a sperm or egg cell with a new, longer reduplication, which may lead to a serious illness in one of their children.

The discovery that speaks to our species' fate came from scientists who compared the DNA of the normal huntingtin gene in humans to the DNA of the comparable gene in several other primates. Of all the primates they looked at, only one species — ours — has a long CGA repeat in its normal huntingtin gene. This long repeat must have appeared by duplication in some common ancestor of our species and the Old World monkeys, but it has been preserved only in the human germ line. We do not know enough about the function of huntingtin protein to say what the purpose of this duplication to our species might be, though it is likely to be significant, since its loss by half in the cells of our brain is sufficient to deplete intelligence and self-consciousness by middle age.

Whatever the reason for the inherited difference among primates in normal huntingtin DNA, it suggests how our species may be unable to maintain its unique self-consciousness indefinitely. Because additional reduplication of the normal human run of CGAs generates a disease that does not show itself until well after the child-bearing years, there is no way for natural selection to select against this repeat. At the same time, the stuttering error of copying never makes the repeat smaller, only larger. Thus, as time goes on, more and more members of our species must suffer dementia.

All I have tried to say in this book can be summarized in one idea: we need to accept that we are all the products of past mistakes. The genetic variations in ancestral species that natural selection chose in order to solve the problem of the survival of our own species were mistakes when they occurred. These ancient mistakes provide us today with, among other things, a brain capable of imagining its own death. Some of the many ways in which past mistakes live on in us are individual, like a mutation in the DNA of a parent; every new case of Huntington's disease is the expression of such a very recent and wholly unavoidable mistake in the human germ line. Other mistakes are more widely shared, like a mutation in a far-distant ancestor, the miscalculations of war, or infection by an inadvertently selected, resistant strain of microbes. Still others, like the normal DNA sequence of a huntingtin gene, are shared by all of us. They are the mixed blessing of our species' birthright.

Scientists and doctors know more clearly than most people that they are made of these genetic variations of the past and that their individual mortality is no more negotiable than that of our species or any other. But knowing this, they too often make one last mistake — an avoidable one, after all — by ignoring their knowledge, suppressing it, and acting as if death, even their own death, were not worth the effort to consider. Attempts to escape from mortality are nothing new, but it is surprising to find medical scientists trying to get away from death by climbing a DNA ladder, even though they know such tricks of the mind cannot work.

This denial of mortality is often accompanied by the denial of another aspect of the human genetic birthright: we are intrinsically social beings. The mind is the product of social interactions; there would not be enough DNA in the world to encode a single mind. From birth on, minds develop in brains by the imitation of other minds, partly but not solely the minds of biological parents. The few behaviors wired into our genes at birth are all designed to maintain and thicken the bonds through which this imitation can proceed. The current biomedical model of a person as an autonomous object lacks a proper respect for these social interactions. It severs the patient from family and social context, and it devalues preventive — social — medicine to an afterthought or a charity.

This denial of the reality of the social bond, like the denial of mortality, is an avoidable mistake of science. These and other strains that have opened between scientific medicine and society are not simply matters of resource allocation. They are signs that the knowledge of death and the need for others in one's life cannot be suppressed any longer, that the dreams of science are no longer satisfying even the dreamers.

In the United States the costs of medical care for eighty-four percent of the people is rapidly closing on a trillion — a million times a million — dollars each year, with no satisfactory national commitment to deliver it to the remaining sixteen percent. It is unlikely that the two intertwined mistakes of today's medical science can be corrected without a renewal of interest in preventive medicine. But what is to be prevented? Prevention has two meanings, depending on what is meant by a healthy person. If health is given a functional definition — you're healthy if you are free to work and think and play to the best of your born abilities — then preventive medicine — in the form of a vaccine, for instance — simply lowers the risk of developing a disease later in life. If, on the other hand, one imagines there is an ideal of human form and function to which we all must aspire, then preventive medicine takes on a different, perhaps alluring, but in the end sinister purpose: the elimination of avoidable deviation from this ideal.

Neither the economic pressure to reduce costs nor the new technologies developed through molecular biology will determine which definition of prevention sets policy. Physicians have already begun to take on the role of gatekeepers, inadvertent agents of selection, eugenicists *manqué*, deciding on the relative value of different human lives. As that becomes more common, definitions of disease will become less a matter of biology than of politics. Half a century ago two books, *Brave New World* and *1984*, envisioned different versions of a future built on the discoveries of science; neither world is one any of us would want to live in.

In Aldous Huxley's *Brave New World*, mothers and fathers have been replaced by "ectogenesis," the fertilization and growth of infants in bottles. Each is nourished and given oxygen according to its future social class, alphas getting the best broth and epsilons a swill that will make them dull but willing servants. A nursery rhyme taught to little alphas and betas goes, "Bye Baby Banting, soon you'll need decanting": Edward Banting was the first to inject purified insulin to treat diabetes, and decanting — the moment when an infant leaves its bottle — is the replacement for birth. Children are taught in their sleep that "a doctor a day keeps the jim-jams away." The "doctor" is the perfect medicine, anodyne, hallucinogen, and tranquilizer, *soma*. As one character in *Brave New World* puts it,

> The world's stable now. People are happy; they get what they want, they never want what they can't get. They're well off;

> they're safe; they're never ill; they're not afraid of death; they're blissfully ignorant of passion and old age; they're plagued with no mothers or fathers; they're so conditioned they practically can't help behaving as they ought to behave. And should anything go wrong, there's soma.

In short, today's dreams of medical science have come true in this brave new world, where death has no dominion.

George Orwell's *1984* is a vision of a future in which the Party is immortal and human lives have little value. While the chance of a future ruled by a totalitarian regime along the lines of Orwell's Oceania may seem quite slim right now, the vision of a suborned and corrupted medicinal science is common to both future worlds, no less useful to a state of drugged bliss than to a state of total war. In both books, medicine is not merely in service to the state but provides its main purpose, "the end to which human beings are to be made the means," as Huxley put it. In both, science and medicine have collaborated with — or created — permanent States in which all sickness, but also all strong feelings and all individuality, are suppressed. In both, the health of a collective society is celebrated as the soundness of a single social body, preserved through science and medicine even though the cells — individual men and women — may change or die.

From the ubiquitous drug soma to the majestic laboratories in which clonal clusters of people are mechanically bred and indoctrinated before birth with the opinions and behavior appropriate to their caste, Huxley's World State is one vast, ongoing medical experiment. Its stratified society is built with poisons poured into the bottles carrying embryos destined to make up the lower orders. Today's technology still falls short of the World State's: we can fertilize a human egg in a dish, but we cannot raise a mouse or a person outside a uterus; we can with precision change a person's mood with a drug, but we cannot tell what a person is thinking. But in medicine many of us gladly accept social stratification reminiscent of the World State, allotting medical care as if some of us were gray-suited alphas and others khaki-jacketed deltas or black epsilons. Our medicine does not approach all diseases with equal panache: the viruses and parasites that largely infect people who cannot pay for treatment are almost entirely untreated, unstudied, even undefined. By failing to make mental health counseling, vaccines, and other tools of preventive medicine as fully and freely available as tobacco, alcohol, and insecticides, we create deltas and epsilons among our children.

On the eighteenth of June, 1940 — the year I was born — Winston Churchill spoke to the House of Commons on the disastrous course of the war of England and France against Germany. He ended with the famous peroration:

> Let us therefore brace ourselves to our duties and so bear ourselves that, if the British Empire and its Commonwealth last for a thousand years, men will say, "This was their finest hour."

Many will recognize this sentence today, but few may recall the earlier phrase it refers to; the reason for "therefore":

> But if we fail, then the whole world, including the United States, including all that we have known and cared for, will sink into the abyss of a new Dark Age made more sinister, and perhaps more protracted, by the lights of perverted science.

What could Churchill have meant in 1940 by "the lights of perverted science"? While he could have been thinking of biological warfare or nuclear bombs, he could also have been referring to the enthusiastic participation by doctors and scientists in the prewar agenda of Hitler's government. This collaboration had already led to the orderly, scientifically planned and executed euthanasia of hundreds of thousands of Germans, and by 1940 these operations had been extended to the East, in occupied Poland.

By the time Churchill spoke, the major organizations of German genetics, biology, anthropology, and medicine and many of the best scientists and physicians in the Reich had spent more than five years in the murder of what the German government and its scientists had agreed to call *Ballastexistenzen*, lives not worth life. Forty-five years after those particular "lights of perverted science" were shut off for good by the Allied victory, the shadow they cast over laboratories and hospitals remains.

In the last third of the century, a new biology has provided medicine with a research agenda and a set of tools and techniques drawn from basic research on the human genome. For every late-onset disease that can be diagnosed before its symptoms appear by a telltale difference in DNA, considerable numbers of healthy people are receiving diagnoses that leave them little to do but wait for the inevitable. The widening gap between diagnosis and treatment has had a second consequence, one that touches one of the most sensitive issues facing us today. Prenatal DNA diagnosis coupled with the termination of pregnancy provides a rational way to avoid bearing a child with a life-threatening inherited disease; more and more diagnoses can be made in a first-trimester fetus, providing a woman with a new and ever-growing set of reasons for the early termination of her pregnancy. But before molecular diagnostic techniques can be properly used on fetal DNA, all interested parties must agree on which versions of any gene are to be considered normal and which are grounds for terminating a pregnancy.

Already the sex of a fetus is a straightforward matter to determine. In the very near future, these techniques will allow pregnant women to decide whether they want to bear a child whose physical and mental states today fall well inside the boundaries of "normal." With time, a combination of DNA hybridization and computer-chip technology may well allow the simultaneous analysis of DNA data on dozens or hundreds of different genes. At that moment, a knowledgeable woman will be able to learn whether her child would be tall or short, hearing or deaf, or gifted or without perfect pitch, or perhaps even gay or straight.

Taken together, these and many other two-edged developments in medical science have defined a new right of privacy: the right to control the information contained in one's own genome. Both law and politics move slowly; the technology is moving much faster than either. As a result, issues of genetic privacy, left to grow in the dark of legal and political neglect, are now able to present us with unexpected and nasty surprises. The "perverted science" of 1940 lasted only five more years before it was brought to a halt by total military defeat. The tools and capacity for its reappearance are, however, nevertheless in our hands today. We need to articulate a vision of the future of science that includes at its center a commitment to acknowledge and honor the personal and emotional content of any experimental research. We need not to manage medical science but to lead it. The critical distinction was well described by the late A. Bartlett Giamatti, the president of Yale and commissioner of Major League Baseball:

> Management is the capacity to handle multiple problems, neutralize various constituencies, motivate personnel. . . . Leadership on the other hand is an essentially moral act, not — as in most management — an essentially protective act. It is the assertion of a vision, not simply the exercise of a style: the moral courage to assert a vision of the institution in the future and the intellectual energy to persuade the community or the culture of the wisdom and validity of the vision. It is to make the vision practicable, and compelling.

Scientists must accept the validity of their own inner voices and see their research as an expression of their innermost feelings.

This will be difficult, but it is not impossible. In a sense, it is just an extension inward of the fundamental methods of science. The great physicist Richard Feynman of Cal Tech saw the possibility as an obligation:

> It is our responsibility as scientists, knowing the great progress which comes from a satisfactory philosophy of ignorance, the great progress which is the fruit of freedom of thought, to proclaim the value of this freedom; to teach how doubt is not to be feared but welcomed and discussed; and to demand this freedom, as our duty to all coming generations.

Today, few scientists accept this obligation, and the public knows it. The popular vision of scientists as white-coated practitioners of a pagan religion is grounded in an understandable uneasiness about the way many scientists present themselves, obeying only their own arcane rules. I first saw the need for scientists to do better when I was an undergraduate. Though I was majoring in physics at Columbia in the late 1950s, I tagged along with my friends to literature classes taught by Lionel Trilling. He was a distant, somewhat foggy creature to me, since he dragged constantly on his cigarettes, and I always wound up at the back of one or another very smoke-filled room. Nevertheless, I knew he was serious about the books he taught, and serious about the world, because an overlap of concerns — the text and the world — marked Trilling's teaching. Even though he sometimes claimed to be interested solely in the words of the text, the world could not keep from informing his interpretations. My colleague Ed-[illegible]ard Said caught this twenty years ago, in this quote from an ar-[illegible] about Said's book *Orientalism:* "In a recent interview [Said] [illegible]ith approval Lionel Trilling's assertion that 'there is a mind [illegible] and argues that it is this mind that the critic should [illegible]doctor, inform, evaluate, criticize, reform.'"
[illegible]d of society" ought to be entirely congen-

ial to scientists. The imagination of a scientist creates a vision of one aspect of the natural world, usually of the world outside the mind, but sometimes an aspect of the mind itself. But that vision is never enough: physical action — experimentation — weighs in immediately to test the model. This back-and-forth of theory and practice works because in science the imagination must either yield to, or encompass, the results of experiment. There is no room in science for empty speculation nor for its complement, the cynical despair we find in so much of today's critical theory.

While the narratives of successful science — discoveries, we call them — are bounded by culture no less than any other narrative, the models they stem from, confirm, and alter are not. These models, the most recently adapted, current working hypotheses of science, float above all their previous narrative versions, persisting through time, never final. We live by such models because they mold the patterns of our thought. In Hamlet's soliloquies, Shakespeare gave us our way of seeing ourselves as having inner voices and developing through inner dialogue. In a similar way, the sciences continue to give us new and sometimes precarious perspectives from which to see ourselves. These, in time, become as completely taken for granted as the Shakespearean notion of a private monologue. In just this way Freud's unconscious and Darwin's natural selection have not merely been added to our vocabulary. They have become aspects of the way we understand ourselves; it is left for scientists to learn that these insights of self-understanding apply to them as well.

I first read Dante's *Inferno* in a general education course at Columbia in 1958. A while ago I returned to it, reading Robert Pinsky's new translation with great pleasure. The *Inferno* is about many things; but to me, it was and still is, above all, an extraordinary example of the power of words to transcend death. Dante meets the damned souls of hell and has the audacity to promise they will have eternal life on Earth if they will allow him to write their stories. They tell him their stories, he writes them out brilliantly, and, after seven hundred years, we still read those stories.

In Canto 31, scientists meet themselves face-on. At the bottom of the last circle Dante sees, in the distance, a Stonehenge of monstrous, missile-like towers. Thinking these to be the Giants of Genesis surrounding the very pit of Hell, he says to us in a parenthetical aside:

> (Nature indeed,
> When she abandoned making these animals,
> Did well to keep such instruments from man;
> Though she does not repent of making whales
> Or elephants, a person who subtly inquires
> Into her ways will find her both discrete
> And just, in her decision: if one confers
> The power of the mind, along with that
> Of immense strength, upon an evil will
> Then people will have no defense from it.)

I have no doubt there will continue to be moments when a science unable to plumb its own unconscious fears and dreams will indeed leave people with no defense from an evil will. It is hard to learn, and then hard to believe, that death is inevitable regardless of the efforts of science. There is a tendency among scientists to respond to this knowledge by withdrawing into themselves, closing the world off by peer review, yet not withdrawal but action is needed. The task that remains, then, is to convert this knowledge into actions of a sensible, honest, and honorable sort, actions consistent with the limitations of science that bring science to its very limits, making it do its best to extend our lives.

APPENDIX

NOTES

FURTHER READING

INDEX

APPENDIX

An Agenda for a More Humane Medical Science

> The honored ideals of the medical profession imply that the responsibility of the physician extends not only to the individual, but also to society, and these responsibilities deserve his [*sic*] interest and participation in activities that have the purpose of improving both the health and the well-being of the individual and the community.
>
> — *American Medical Association Principles of Medical Ethics* (1957), Section 10

Would biology or medical science change in any important way if this book were to be taken to heart by those now making the decisions that will emerge as the research policies of the early twenty-first century? While we wait to see, here are a few specific suggestions to start things off in the late twentieth.

1

Microbes do not respect national boundaries; the strongest ally infectious agents have is the human notion of national sovereignty. International cooperation was a prerequisite to the elimination of smallpox. If every person on the planet could simply be vaccinated with the vaccines we already have, hundreds of millions of people, a good fraction of them babies, would be saved from dying.

Only a few agents of infectious disease — yellow fever, an insect-borne virus; Llasa, viral hemorrhagic disease; smallpox; cholera; diphtheria; tuberculosis; and plague — cause illnesses that must be reported to the United States government today. All others, including malaria and all antibiotic-resistant strains of common infectious microbes, come and go unremarked. Many other diseases used to be reported; the shortsighted decision to save a small amount of CDC money guaranteed the fast and distant spread of any outbreak of antibiotic-resistant infection. It also mistakenly presumed that the United States had no need to worry about tropical diseases like malaria, even though the climate of the southeastern United States would suit the insect vector quite well.

To pay for a more rational and comprehensive defense against microbes, we might consider using a version of the military model which is not based on a fantasy of total victory. There is a pleasing symmetry to extending the notion of subsidy for the sake of security from the production and purchase of lethal weapons to the production and distribution of life-saving vaccines. The underlying logic of a military model for mobilization to assist our immune systems through vaccines is the opposite of the SDI notion of perfection. Instead of SDI we need an SVI — a Strategic Vaccine Initiative. SVI would acknowledge that our best hope is a standoff and that our best strategy is to help our immune systems turn microbial mutability to our advantage by domesticating the microbes that get inside us.

In contrast to SDI, SVI could work only if it were the product of total international cooperation. Political, religious, and ideological differences make no difference to tuberculosis or malaria; they have no place in species-wide SVI. National sovereignty may seem an impermeable barrier to the necessary transnational attitudes and actions, but we have a precedent at our fingertips for the permeability of national borders to new technologies. Ideas and information that get onto the Internet travel around the planet, crossing national boundaries with impunity. Organized and run from the beginning on the Internet, an internationally funded SVI would not need to have a single location in any one nation. That would be an appropriate organizational strategy for the kind of international effort it will take to respond — as a species — to the invisible species that will always threaten us. Like the immune system in any of our bodies, the Internet is widely distributed, rapidly adaptable, and quick to learn. A new idea that travels through the Web is quite like a new antigen that stimulates a strong immune response. And like the chemicals and cells in a person's immune system, ideas that move through the Web may be what keep our species going, especially if one or more of the microbes we live among gets going in us in a serious new way.

2

The ideal vaccine for any infectious agent should be safe, oral, and effective when given in few doses early in life. The new technologies and insights of molecular biology can and should be brought to the task of creating such vaccines. Only twenty or so vaccines are available in clinics today. Bringing any of them closer to this ideal would be a way to save a lot of young lives. One promising line of work takes off from the way that foreign molecules are recognized: the immune system is tuned to detect foreign bodies best when they are presented on the surface of the body's own cells. The immune system responds to protect the body at the expense of some of its cells: first it destroys the cell that presents the foreign molecule, then it fills the bloodstream with chemicals — antibodies — to find and signal the presence of additional foreign material elsewhere in the body.

Live vaccines stimulate this entire response best, which is why their effects tend to last longer than those of simple chemical vaccines. But one chemical — DNA — has the potential to stimulate the entire immune response in complete safety. When DNA molecules encoding the information for the production of foreign molecules are injected under the skin, some are taken into immune cells. When these cells express their new gene, the rest of the immune system attacks the "infected" cell and pours out antibodies to the DNA-encoded molecule. Today's experimental DNA vaccines tend to stimulate real but rather weak immune responses. Since it is both rational and safe, this technology is poised to take off once vaccines are once again recognized as the most practical way to deal with infectious diseases of all sorts.

The simplest way to revisit Jenner's work, in light of our new capacity to move genes around, is to engineer attenuated microbes that protect against pathogenic ones, recreating by design what Jenner found by serendipity: that infection by cowpox could protect against future infection by smallpox. Attenuating a microbe means modifying its genome so that it cannot mount a full infection; instead it becomes the biological vector for an immunizing stretch of pathogen-specific DNA. We may yet elevate the vaccinia virus to just the state of ubiquitous presence in our bodies that its distant relative smallpox so earnestly aspired to reach. The full DNA sequence of vaccinia's genome has been known for almost a decade; now that smallpox is done for, many scientists have begun to turn to the vaccinia genome as a recipient for new genes, hoping to make the virus into an all-purpose vaccine against any number of other pathogens. Oral polio virus vaccine has also been engineered to express the genes of many other viruses. Most of these engineered vaccines are still experimental, although human adenovirus, engineered to carry genes from the hepatitis B virus and to infect but not to kill human cells, is used by the United States military as a hepatitis vaccine.

3

Oral vaccines available today are prepared from infected, cultured cells. Although it is attenuated, the Sabin live polio vaccine can be taken by mouth because it can still infect the lining of the intestines. It is safe because its genome differs from the pathogenic polio virus in enough places to assure that it will not revert to its ancestral capacity to go from there into neural cells. An oral vaccine is by definition edible; another way to make an oral vaccine would be to put a few of a pathogen's genes into the germ line of an edible plant, forcing offspring plants to produce antigenic foreign proteins and thereby make them into edible, even nutritious vaccines. Transgenic plants are now being tested for their ability to serve as cheap, stable oral vaccines against hepatitis and cholera. The main limitation so far seems to be tolerance: the intestinal immune cells that see the foreign protein as part of a digested mass of plant material cannot tell that it is foreign unless it comes in as part of a larger, more obviously microbe-like structure.

4

Most current drugs slow the progress of AIDS by interfering with the virus's ability to copy itself early in infection and to establish an infection in a new cell. Once T-cells are compromised, it is still possible to slow the growth of the virus, and thereby to slow the course of the disease, with drugs that inhibit a viral protein called protease. Protease inhibitors have serious side effects, and once initiated, a course of treatment with them cannot be stopped without the T-cell damage getting worse. Although many people are alive today who would not have survived without them, it is not clear that a person ever fully recovers from AIDS by taking any of these drugs. The domestication of HIV remains a largely untested alternative to current proteases and other antibiotics. Drugs that would specifically block the interaction between any new HIV variants and T-cells would not select against the growth of the original HIV in macrophages, so they ought not to select for drug-resistant strains. If HIV could be domesticated inside the macrophages indefinitely this way, it would leave someone HIV-positive but relatively symptom-free.

5

If the ideal preventive medicine for infectious diseases would be the delivery of an optimal vaccine for all diseases, a solution at hand comes pretty close: mother's milk. A nursing baby winds up with a fiftyfold enrichment of its mother's immune-protective molecules. Milk also carries natural drugs to fight infection, in particular the anti-inflammatory agent lactoferrin and the antibiotic lactoferritin, as well as sugars that trick bacteria into binding to them rather than to the surfaces of a baby's cells. A baby's immune system is set for life by the mother's milk: an organ transplanted from mother to child will take with much greater ease if the child has been breast-fed.

A complete response to microbial disease must begin with the commitment to encourage and assist every mother to nurse her newborn child before it is exposed to any vaccines, let alone any antibiotics. Breast-feeding so enhances the immune system that cultures that do not breast-feed have a tenfold excess of infant mortality over those that do. This difference is due to the absence of similar enhancers of the immune response in any other foods and to the relative contamination of all foods compared to milk from the breast, which is sterilized by the mother's immune system. In a 1994 editorial in the *Journal of the American Medical Association*, two physicians from the CDC described the worst consequence of our having devalued vaccination in an age of artificial formulas and antibiotics: these have helped infectious disease to get back into our children.

6

A cancer prevention agenda for basic research would begin with a planetary review of differences in the incidence of various cancers, because some regions and cultures are hot spots for some cancers, while in others the same cancers are exceedingly rare. From this international effort, governments and companies worldwide would have the information necessary to plan a planetary strategy for the prevention of cancer: planetwide optima for low-mutagen food, air, and water and clear guidelines for behaviors that would, together, assure the lowest possible frequency of avoidable cancers. In this context, the current emphasis on the genes responsible for a tumor would be seen for what it is: an interesting sidelight to the real problem of cancer, not the main issue.

At present, we search for populations at high risk for inherited cancers only to tell families what their fates will be. We spend relatively little time and money understanding the origins and consequences of the habits that bring on the majority of fatal cancers and reaching out to the entire population with help in avoiding these habits. A 1996 study by the Harvard School of Public Health found that only about ten percent of people who had died of cancer were born with versions of genes that made the disease inevitable. About seventy percent of the lethal cancers were brought on by choices such as smoking, poor diet, and obesity, and most of the remaining twenty percent could be attributed to alcohol, workplace carcinogens, and infectious agents. Smoking is optional, but eating, drinking, and breathing are not: the task of understanding why people act against their own best interest even after they learn how to act prudently is not part of today's agenda for cancer research, but it should be.

7

Setting prevention aside — not because it is impossible, but because in scientific terms it is so easy that one is embarrassed to say more about it — in the near future cancers are likely to be dealt with by a slowly evolving combination of genetic, immunologic, and antibiotic interventions. The lessons of microbial research apply here: the immune system is the body's first line of defense. One genetic strategy aims to restore the function of the gene that has mutated in a clone of cancer cells, thereby redifferentiating the clone so that it no longer grows. Tumor cells that redifferentiate will eventually cease to grow and spread without needing any further treatment.

Reports have begun to appear of drugs that renormalize tumor cells and of others designed to force tumor cells to commit suicide. The simplest use of DNA-based information would be to undertake a DNA-based treatment of the problem. If the activity of BRCA1 or BRCA2 were somehow to be returned to the cells of a breast tumor, they ought to revert to quiescence, curing the disease without the side effects of current treatments. However, there is a catch. Most growth-controlling genes work through proteins that switch other genes on or off. These proteins never leave the nuclear sanctum of the cell they keep quiescent. Any drug designed to mimic such a protein would have to get to the tumor cells — every last one of them — get inside each, get to each nucleus, and find the same set of other genes to turn on or off. This seems unlikely, and in fact to date no laboratory has been able to mimic the effect of an absent tumor-suppressing gene except by introducing the gene itself into a tumor cell, a trick unlikely to work in a clinical setting, where even one untreated tumor cell would be able to seed a brand-new tumor.

Coupled with a real commitment to prevention, these new classes of drug promise us a real hope of dramatically reducing the incidence and the damage of cancer. One promising line of work builds on the absence of a functional version of the p53 gene in most tumors. Scientists have assembled a recombinant version of a human respiratory tract virus called adenovirus, which cannot grow in the presence of p53 protein. When the virus is injected into immune-deficient mice bearing human tumors, the tumor cells are the only cells to die; preliminary reports on patients with late-stage, p53-less tumors are promising, but the technique is far from general applicability. There are drawbacks to be expected: for instance, a cancer patient with a good immune system is likely to defeat the treatment, and any treatments to inhibit the immune system may lead both to an overgrowth of the tumor and to the appearance of aggressive new infections.

Dolly, the cloned sheep, and others like her may be useful in developing new treatments for cancer: the cytoplasm of a mammalian egg is the only place we know of today that will direct the expression of each differentiated pathway necessary to the construction of all our tissues. Adding a nucleus from a cancer patient to a fertilized egg cytoplasm, it might be possible to grow any differentiated tissue in a dish and have it be wholly acceptable to the donor. In this way it might be possible to replace a tissue like the liver after excising the original to rid the body of all traces of a liver tumor. More generally, it ought to be possible to rebuild a person's immune system in a dish this way, and even to stimulate it in advance, to attack the pathogen that is attacking the body, whether microbe or tumor.

Some diseases — alkaptonuria, for example, or cystic fibrosis — are the consequence of a protein's complete absence from the body. Getting the correct gene into even a few cells can mean at least partial success at ameliorating such diseases. In at least one case — the absence of an immune system due to the inherited lack of a protein called ADA — a cure has been at least temporarily achieved. After a decade of living in a germ-free isolation tent, a child suffering from ADA deficiency — whose blood cells had been removed, given DNA encoding the gene, and transfused back — developed a functioning immune system and has been able to leave the tent. Gene replacement has been touted as a new therapy for cancer, but there is reason to be skeptical: a decade after the first published trials and despite a few thousand reported attempts, no gene therapy has yet reached the status of a standard treatment for any disease.

8

Finally, here is a suggestion that an increasingly popular line of work be avoided.

DNA transfer into embryo cells, the line of work that gave rise to Dolly, has tempted some scientists to contemplate a radical solution to the problem of an inherited propensity to develop tumors: implant the missing growth-control gene in the cells of an early human embryo so that a person is born with a functioning version of the gene in all tissues. The procedure is called germ-line gene therapy. It is a hybrid combining the laboratory technology of delivering DNA directly to a tissue that lacks a necessary gene with the clinical skills to bring sperm and egg together in a dish, allowing couples to have a baby when their own bodies cannot support fertilization of the ordinary sort. In vitro fertilization combined with gene therapy works well in animals. Any DNA that successfully inserts itself into a cell of an early embryo will be copied into all the descendants of that cell, so treated embryos will grow into animals born with a new gene in at least some cells of all their tissues. Under conditions where the switches to turn on the gene are also in their proper place, a new gene makes a normal amount of functional protein in just the cells that would normally have it, so that an animal that would otherwise suffer from an inborn genetic defect can be quite normal at birth and live a normal life thereafter. And, since the implanted DNA is copied into the cells that make sperm or eggs — the germ line — such an animal's offspring will also be cured of the genetic illness forever.

In the conservative sort of germ-line gene therapy, a couple susceptible to a disease because of an inherited damaged version of a gene would donate sperm and eggs to a laboratory. The fertilized egg would be allowed to grow for a while in a dish, then it would be injected with the DNA encoding the missing gene, slipped into the mother's womb, and left to grow into a child with, one hopes, a good copy of the gene functioning in each cell. Injecting a growth-control gene into the early embryo formed from sperm and eggs donated by a couple from susceptible families would not be enough, since a child born with even a few cells lacking the injected gene would be at risk for cancer developing in those cells. Instead, germ-line gene therapy for cancer would have to succeed in putting the growth-control gene into every cell of an embryo. This could be done, but only by inserting the growth-control gene either into the single fertilized egg cell before it divides or into the germ cells of both parents. Along these lines, a few laboratories have recently reported the successful addition of a foreign gene to the chromosomes of mouse testicle cells growing in a dish. When these cells were returned to the testicle of a mouse sterilized by irradiation, they produced functional sperm, making the male mouse a donor of the foreign gene to all his offspring.

Germ-line modification has the elegance of a complete solution. Unfortunately, it is a solution that sacrifices the current generation for the next, and as such it does not serve the purpose of medicine, that is, to alleviate or cure the suffering of a person already here among us. Neither parent of the modified embryo is in any way treated by the technology that would introduce a DNA into an embryo's cells; afflicted parents would not be cured, nor would parents at risk of cancer, say, be any the less at risk if they agreed to let their embryos experiment. Instead, at best a baby would be born with no more but no less risk of cancer than anyone else. At worst, the child might be born with another, unrelated form of inherited disease, the result of an inadvertent mutation caused by the insertion of the DNA into one of its cells.

The creation of any child with a changed genome would be a Promethean grasp at the human germ line. It would also be an act of enormous hubris, risking inadvertent chromosomal damage that might not show itself in the growing child for many years. This line of research has already raised some new social and legal problems. It has obliged us to decide if we are willing to pay the price of converting kinship and childhood into commodities in order to find out whether these techniques will work properly; it has given us the task — as yet unfulfilled — of setting a proper boundary on the freedom to initiate genetic novelties in our own species.

Notes

Introduction

1. King Solomon's wisdom was first recorded in Hebrew; I have used *Tanakh*, the 1985 Jewish Publication Society's translation according to the traditional Hebrew text. The King James version, though more familiar to many readers, is not as close a translation of the Hebrew: "Then spake the woman whose the living child was unto the king, for her bowels yearned upon her son, and she said O my lord, give her the living child and in no wise slay it. But the other said, Let it be neither mine nor thine, but divide it. Then the king answered and said, Give her the living child, and in no wise slay it: she is the mother thereof."
2. Up against this limit, even time itself yields: the theory of relativity tells us that as objects are pushed to approach it, they appear to themselves to get heavier, while time appears to them to slow down. We can make time slow down for a few atoms by pushing them to speeds approaching the absolute limit of the speed of light, but it takes a lot of energy and a tool the size of a cathedral — a superconducting supercollider — to get them going fast enough to measurably slow their time. Nothing alive is small enough to have time be slowed this way. Nevertheless, energy permitting, the oddity of time's malleability would apply as well to the earthbound living world of which we are a part. For instance, if one of a pair of twins were to be launched on a looping trip around the nearest star at a speed near that of light, the round-trip excursion would appear to the earthbound twin

to take about a decade, but the lightning-fast twin would experience only a relatively short time away and return far younger than the one who never left.

1. Sensation

1. Large creatures built from many cells — each the descendant of one fertilized cell — are a rather new event in the history of life. Until about a billion years ago, even the most complex living creatures were all one-celled. Though they thrived in the most diverse of environments, they all looked more or less like today's bacteria and algae, blanketing the earth's shorelines with an ever-thicker scum of disconnected cells. Though of recent origin, the developmental clock, which makes the descendants of a fertilized egg into a person rather than an accretion of the same sort of scum, does this by differentiating cells of an older, original design, one that first appeared some three billion years ago and has been shared by all cells ever since. The genes in the nuclei of our cells are the direct descendants of the genes of these ancient cells, still made of DNA, still copying and retaining information and passing it on from generation to generation just as they did. To turn genes on and off in response to the needs of the moment, our cells turn to the same sorts of molecules — proteins and hormones — that those ancient cells used for the same purpose. The developmental clock is just a newer use of this original machinery, turning on wholly different sets of genes in different cells, thereby creating the different tissues of the body from the genetic descendants of a single fertilized-egg cell.
2. Nerve cells tell time at their tips as they release and take up small molecules called neurotransmitters. To help the recipient cell mark the time it receives a signal to within a thousandth of a second, the tip of the sending nerve cell vacuums up any leftover molecules of transmitter that have not made it over to the receptors on the receiving cell within a thousandth of a second of

their release. Many psychoactive drugs inhibit a sending nerve cell's reuptake machinery. Prozac, for instance, is an inhibitor of the reuptake transporter of serotonin, and cocaine blocks the dopamine reuptake transporter. Such drugs increase the likelihood that a set of connections that use a particular neurotransmitter will fire successfully, but they do this at the cost of blurring the instant at which the firing will occur. The fact that transmitter reuptake inhibitors can strongly affect such aspects of consciousness as mood and anxiety is certainly consistent with the notion that consciousness itself is a brain state maintained by proper and precise timing.

Mice that have been genetically modified to lack a functional dopamine reuptake transporter are permanently hyperactive but wholly indifferent to the administration of large doses of cocaine. Parkinson's syndrome is a disease of the brain and body that results from a failure of portions of the brain to produce enough dopamine to stabilize certain neural networks. It would be interesting to find out whether cocaine alleviates the Parkinsonian symptoms of tremor by allowing whatever little dopamine that makes it to the synapse to sit around for a longer time. Such a study would, however, be difficult to justify on ethical grounds, considering the highly addictive nature of the drug.

3. All light — the light by which we see, the light by which our dentist X-rays our teeth, and the dwindling light by which we know the universe had a singular beginning — is simultaneously a set of waves of electric and magnetic energy traveling at the invariant speed of light and a set of dimensionlessly small points or packets, called photons, traveling at the same speed. The particle-like and wavelike facets of light are two sides of one coin, a duality no more separable than the Trinity is to a believing Catholic. The way in which light is looked at will determine whether its wavelike or particle-like properties will dominate. If light passes through two slits, its waves interfere to form an image of fringes on a screen. Conversely, as Einstein first explained, if light hits a metal surface, even a few photons

of sufficient energy — that is, light of the right color — will displace a photoelectric current of electrons, while light the color of less energetic photons, no matter how bright, will not.

4. The similarities of a TV to the eye and brain is a good example of technical evolution driven by biological evolution: the technology of TV mimics the arbitrary choices of color differentiation made millions of years ago by natural selection, working on our distant ancestors. To transmit a scene so that the dots of a monitor glow in the right way to convey its colors, a television camera passes it through red, green, and blue filters, creating versions quite like those initially picked up by the three different sets of cones in the eye. Electronic circuits then play the role of blind retinal cells, converting these three monochromatic scenes into three abstract ones for transmission, albeit not to the brain, but to a humble TV set. One abstract signal simply adds up the three images for a measure of overall brightness at all points; it resembles the output of additive retinal cells. The other two are quite like the older and newer subtractive circuits in the retina. One of these different images represents blue as bright and yellow as dark; the other represents red as bright and green as dark.
5. The light-absorbing proteins in these two kinds of cones are very similar to each other: the long-wavelength protein absorbs yellow-orange light best, while the medium-wavelength protein is most sensitive to a greenish-yellow wavelength only one tenth of an octave shorter. Yoked together through retinal connections, the two cones can pull most of the colors we see — all colors from red through yellow to green — out of less than half the visible spectrum.

2. Consciousness

1. Early assays of brain activity — as opposed to behavior, or the output of brain activity through bodily responses — were limited to the detection of an average electrical activity, a mixture of waves from the brain. The name of the tool that recorded these waves, an electroencephalograph (EEG), was rather more com-

plicated than its output, a strip of paper covered with wiggles representing the total electrical activity coming through the skin at various places on the skull. EEG records are gross averages, as blurry a representation of the brain's activities as the crowd noise at a baseball game is of the conversations between any two people. EEGs could be coaxed into giving information about the normal brain's way of responding to a specific stimulus by asking the subject hooked up to the machine to look at a flash of light or to listen to a short tone over and over again. Each EEG trace showed a blip correlating with the reiterated event. Since the subject could not control EEG responses — neither by thinking about the repeated stimuli nor by ignoring them — whatever the blip captured of the brain's activities was not related to conscious mental activity.

Certain novel variations in a repeated stimulus — one new tone in a long series of identical tones or one flash of a different color — did catch a subject's attention, but even these novel events did not change the size or shape of the blip on the subject's EEG. This result seemed to confirm the distinction between behavior and mental activity, supporting the notion that neither conscious nor unconscious thought was a real event in the brain. This approach to the problem of the mind had the virtue of ending the problem for those who accepted it.

2. The first sign of conscious mental activity emerged from computers that could merge and average the EEG signals from the brain of a person subjected to many thousands of repeats of the same stimulus. When a subject was unexpectedly presented with an oddball version of an otherwise stereotypically repeated stimulus, a new blip — an event-related potential (ERP) — would appear about one third of a second after the novelty. An ERP always followed a rare yellow light in a string of blue ones, or a single B-flat in a string of Cs. Usually, subjects noticed the difference that stimulated an ERP; in those cases the ERP was a physical sign of the brain's mental work in updating its internal representation of what was going on in the world.

ERPs represented changes in the brain that were part of consciousness; once consciousness had been made experimentally accessible, it became real even according to the strictest scientific construction of reality. On the other hand, an event reflected in an ERP is not quite an internal monologue, a feeling, a memory, nor a remembered dream. ERPs could not be markers of even the simplest introspection, since they needed to be evoked by external stimuli. An ERP might be a sign that perceptual consciousness — the state of being aware of something — involved real and detectable changes in the brain, but it contributed hardly anything to the question of whether reflective, introspective consciousness — the recognition by the thinking subject of his or her own acts or feelings — could be studied as a set of brain events. Nor were ERPs easily localized to specific sets of molecules, circuits, or places in the brain; even hitched to a computer-driven display, EEGs could not resolve events very well to specific places in the brain, nor to specific instances briefer than a tenth of a second.

The second wave of computer-based technological change, which swept these problems aside, came with the development of dynamically changing images of the inside of the brain, pictures that showed the brain engaged in mental work, thinking as well as reacting. Volunteers willing to be injected with a mildly radioactive isotope that emits the unintuitive positron — an electron with a positive rather than negative charge — allowed these instruments to visualize the electrically active places in their brains by the large amounts of sugar these regions draw from the bloodstream. Images of the brain's electrical activity generated this way are called positron-emission tomographs or, more informally, PET scans. A PET scan reconstructs events inside the whole brain, and by showing where extra sugar-laden blood is going, it can reveal which brain circuits are markedly more active, producing blurry movies of a changing brain.

A new sort of scan, magnetic resonance imaging (MRI), did not detect brain function but could reveal structural details of the brain as small as the head of a pin. PET and MRI scans obtained under controlled conditions were soon merged to correlate electrical activity in precise locations in the brain with specific behaviors and mental activities. In this way, characteristic and reproducible changes in the brain were seen to occur as a person's perceptions changed. When tested subjects heard or saw something, and then later when they thought about what they had seen or heard, the brain changed; remarkably, changes in the brain that accompanied the thought of a picture or a melody were all but identical to the ones that accompanied the experience itself. These experiments allowed the analysis of mental states as brain states, but they offered no mechanisms to explain the brain activities they uncovered nor any idea of how those activities might be connected to the subjective sense a person has of being conscious.

3. The thalamic clusters are linked by unusual connections that inhibit each other's output signals as well as ordinary, stimulatory connections. The inhibitory connections keep the hum volume stable, preventing the ever-stronger mutual stimulation that causes an audio system to howl when the speaker's output is picked up by the microphone; epilepsy is often the consequence of such runaway mutual stimulation in other regions of the brain.

4. A few years ago I experienced the difference when I began to study Hebrew for the first time. Its alphabet of written consonants, pointed with dots and dashes to signify syllables and vowel sounds, was not too hard to pick up. But if I must read a completely unfamiliar Hebrew passage aloud, I find I have to spend a certain minimal amount of time bringing each syllable to consciousness, then deciding to voice each successive Hebrew phoneme. If I read aloud slowly enough — it seems very slow to anyone comfortable with Hebrew — I am fine, but if I try to speed up even a bit, I jam up completely. Many forms of learning impairment in children correlate with an inability to produce or perceive differences in sounds in a single sweep of the cortical binding clock. This impairment is sometimes the consequence of

a defect in the thalamus. For example, language-impaired children who typically cannot distinguish changes in tone unless they occur no more than ten times a second often show a reduction in the amplitude of the forty-cycle-per-second thalamic signal.

3. Memory and the Unconscious

1. The first psychoanalysts did not have a clue about what mechanisms might underlie the repression of unconscious memories, nor their emergence as dreams or fantasies. Initially they proposed that the adaptive value of repression lay in its ability to minimize the irritation of particular feelings from sexual and erotic desires, as if there were an emotional thermostat that sought a mental Nirvana, free of anxiety and sexual frustration. In his later years, Freud turned to the darker notion that the inevitability of death gave an intrinsic coloration to our memories and behaviors, and that our minds repress memories to relieve the tension between urges for the peacefulness of Nirvana and the silence of the grave.

More recent attempts to place repression in a context of natural selection have argued that in order to become social creatures, we needed evolved minds that require and actively seek a wide range of emotionally gratifying human contacts. This awkwardly named "object relations" model sees the division of the mind into conscious and unconscious portions as having evolved to enhance the chances of concerted, focused attention on human interactions. Dreams, creative acts, play, daydreams, and fantasies — fulfillments to the early analysts of the wish to turn away from reality — are in this view of the mind adaptive in their own right and functional in the present moment, enabling a person to mentally sort through possible new relationships, matching up inner and outer experiences. Clinical observation of very young babies and their parents supports at least one prediction from the object relations model: emotionally rich but humane interaction with a loving adult is as important to an infant as food or water.

2. Once our ancestors learned to stand upright, sight and sound rather than smell set the social mood of the moment. Perfumes and dry underarms do remain serious business, though, because the ancient anatomy of our brain still associates odor more than sight or sound with the emotional affects of connection with — or separation from — another person.
3. These examples of lesions in the brain and their effects on the ability to dream are taken from the notes of an unpublished lecture given by the distinguished British psychiatrist David Solms to the New York Psychoanalytic Association in 1995.
4. Baseball is another game built on a myth, one also shared by societies from Homer's to our own: that there is no place like home. Each side has an equal number of chances to leave home and return; getting home often enough is all it takes to win, and winning is everything.
5. The Romans, always happy to add a god to their Pantheon when it might make things better for them, elevated Asklepios to the status of a god during a plague in 293 B.C.E. on the instructions of an oracle.
6. The experiments that led to Jacob's Nobel Prize revealed how cells chose which genes to open and read, converting their information from the data tape of DNA into the three-dimensional, sculptured machines we call proteins. Kornberg's experiments elucidated the mechanism of one set of proteins, the ones responsible for copying DNA.

4. The Fear of Invasion

1. Tuberculosis can be treated with antibiotics today, but there are still five million new infections each year. These bacteria hide so well from the body's immune system that most people harboring a few somewhere in their lungs have no clue that they are infected until an intentional scratch with an extract of the bacteria

raises an itchy red bump, a sign that their immune system has a memory of the infection. Once an infected person's immune system loses potency — most commonly because of old age, malnutrition, or a secondary infection by the AIDS virus, HIV — tuberculosis bacteria rapidly proliferate and enter their contagious phase, converting lung tissue into bloody nodular nests. A cough will then send the bacterium on its way, from an infected lung to an uninfected one nearby.

Past a certain point, the lung nodules of emergent tuberculosis bacteria are lethal; the mortality rate of active tuberculosis approaches fifty percent in untreated people; it is higher in children. For every million people infected each year, another half-million or so — infected years or decades earlier — will die of an exploding time bomb of bacterial growth.

2. One unintended consequence of smallpox's obliteration may have been a worldwide increase in the incidence of asthma. When vaccines reduce childhood infection rates, childhood asthma often increases. This is a measure of the seriousness and the duration of our struggle with microbes. Asthma is an excessive and dangerous immune response in the lungs and bronchial tubes, stimulated by breathing in an ordinarily innocent molecule. In the first three months after a child is born, its new immune system, poised to face the world, has to respond not only to infective agents but also to myriad new proteins that reach it with every breath. Until very recently, infections in that period were unavoidable, so the immune system was selected to expect them. When a child born today is exposed to an infectious agent in that period — whether inadvertently or through vaccination — its new immune system begins to develop in ways that allow it to work optimally in later life to head off other infections. The smallpox vaccine used to serve as one certain infection. Since smallpox was obliterated, no one is vaccinated any longer. An unvaccinated infant's inexperienced immune system may become acutely sensitive to nonliving ambient foreign molecules instead; in other words, the child may develop asthma.

3. Globally, pesticides are dispensed for various purposes at the annual rate of about one hundred grams — three ounces — per person. Only about a tenth of that is intended to control the insects that carry human disease. The rest is used on crops, to control plant pests. Using pesticides on crops will continue to select for pesticide-resistant insects, some of which undoubtedly will, like the mosquito, carry a human pathogen.

   People living in crowded places tend to create habitats that mosquitoes use, as if we and they were drawn to each other. The old car and truck tires that litter any terrain once occupied by a vehicular society, for instance, are notoriously hard to empty after a rain, so they turn out to be wonderful incubators for mosquitoes in precisely those places where malaria is most likely. Global warming is a seemingly inexorable aspect of carbon-fueled intensive industrial development. It is caused by the cumulative effect of pollutants like carbon dioxide trapping heat inside our atmosphere. The warmer the world, the more stagnant pools of water reach the right temperature for mosquito larvae to hatch; even small increases in average temperature favor the spread of mosquitoes and other warm-weather pests to currently temperate portions of the globe.

4. Attempts to develop an HIV vaccine have focused on its fastidiousness. HIV uses molecular signals to find and enter just one set of immune cells, called T4+ helper cells. When it does hit one of these cells, the virus sticks to a special pair of proteins found only on their cells' surface, and from there it penetrates the cell with a syringelike protein on its own surface. One of the two specific proteins HIV needs to find, called CD4, is also the protein that makes an immune cell into a T4+ cell, so its role as an HIV receptor seemed both necessary and sufficient to explain HIV's preference for T4+ cells. The second receptor, called CCR5, was unexpected. It is not restricted to T4+ cells; it is the receptor on many kinds of immune cells for a family of small molecules that regulate the immune system, called chemokines. The fact that HIV uses a second receptor explains an early failure in HIV vaccine development: vaccines that immunized

against the part of the virus that recognized the CD4 receptor did not work well. Vaccines are currently under development that will immunize against the parts of the virus that recognize the CCR5 receptor as well as the CD4 receptor; these ought to block HIV infection.

When HIV takes over an immune system and destroys it, all vaccines necessarily lose their force. In some cases, that can be doubly dangerous. For instance, nonspecific, spillover immunity from the smallpox vaccine is sufficient to protect people from infection by a close relative of smallpox, the monkeypox virus. Thanks to its complete success at eliminating smallpox, the vaccine is no longer being administered. However, a return of monkeypox disease has been one unintended consequence in regions of Africa where people live close to other primates. In the ordinary course of events, smallpox vaccine would quickly fix the problem, but in areas where HIV is epidemic, smallpox vaccine cannot be administered because HIV-infected people have such weak immune systems that the live vaccine simply kills them. Vaccination is a strategy that depends on a sound immune system; monkeypox can no longer be stopped by vaccine in HIV-positive parts of Africa.

5. Currently more than a million American children under the age of three have not received the vaccines that would have immunized them against diphtheria, pertussis, tetanus, polio, measles, mumps, and rubella. Each of these children is a walking time bomb. It would be easy to write off the problem as one of ignorance and poverty, and it is that, but it would be wrong to presume that people with few resources are simply bad parents. The problem often as not is a doctor's ignorance and the poverty of our vision for the future. Many of these unimmunized babies have been seen by their doctors more than once but have simply not been vaccinated. Whatever other reasons there might be for the embarrassing deficiency of childhood vaccination in current medical practice, professional amnesia — or denial — is the main cause. The physicians surveyed thought ninety percent

of their patients had been vaccinated when fewer than forty percent actually had been.

5. The Fear of Insurrection

1. The scientific study of the human germ line — human genetics — began as the last century ended in the clinic of a young London physician named Archibald Garrod. He saw a few patients with an odd problem: their urine turned black after a few hours. Some of them were infants, brought in by parents quite upset at the sight of black nappies. Garrod checked with his colleagues throughout London and found a few dozen other cases. The syndrome was called alkaptonuria, after the name of the chemical that turned black. What caught Garrod's eye was not the chemistry but the remarkable persistence of the difference between people with and without black urine. A person either secreted a lot of the chemical, a few grams each day, or the urine showed no trace of it; no one made an intermediate amount, and no one's status changed, even over a lifetime.

In Garrod's search for people with black urine, he also noticed another unexpected constant: half or more were the offspring of first cousins. As he pointed out, this was odd because although there were about fifty thousand such children in London, only six had black urine: how could consanguineous parentage be so necessary and yet so insufficient? With a great intellectual leap he surmised that the black-diapered offspring of cousins were no mere oddities but rather signs of the operation of Mendel's laws of inheritance in human families. Mendel had shown in pea plants that some parental differences were passed to the next generation as if they were discrete alternatives, with each parent plant contributing one alternative to each offspring plant.

Mendel also showed that alternative states were inherited according to one of two patterns, which he called dominant and recessive. If a plant received one from one parent and another from the other parent, the offspring would resemble the dominant parent; the recessive alternative would not appear unless the offspring received copies of it from both parents. Garrod guessed that black urine was inherited in a recessive way and that it resulted from the inheritance from both parents of copies of the same variant version of a single trait or, as we would say, gene. That would explain the relation of the black urine to consanguinity: first cousins share a common pair of grandparents. If one of those grandparents were to have a single copy of a variant gene associated with black urine, neither he nor any of his descendants would have a chance of expressing the trait; but if two of his grandchildren were to marry and have children, each great-grandchild would have a good chance of sharing the same variant gene from its great-grandparent and of thereby having black urine from birth.

In allowing the possibility that humans inherited their differences in the same Mendelian manner as pea plants and pigeons, and in then concluding from this single example that a clear and singular chemical difference — inherited in the pattern predicted for a single variant gene — could be the cause of a human difference, Garrod in one sweeping insight founded human genetics. When later work showed that alkaptonuria was not — as he had thought — simply an "alternative course of metabolism, harmless and usually congenital and lifelong" like albinism, but a debilitating, life-shortening arthritis from pigment roughening the articulating surfaces of various joints, Garrod's surmise became the foundation of genetic medicine as well.

Because alkaptonuria is rare, affecting only one person in a quarter million, it has taken a century for the mutation that causes it to be found among the tens of thousands of genes in the human genome. The gene was recovered from human chromosomes in 1996, not by physicians, but by scientists working on the genetics of bread molds. In a stunning integration of Mendel with Darwin, a Spanish laboratory used Darwin's insight —

that all current life is related through common ancestors — to trap the human gene involved in alkaptonuria by its similarity to a gene responsible for the formation of a similar black pigment in one of their bread molds. The human and bread mold genes encode similar proteins; each is a different current version of what was an ancestral protein in some unrecognizable, now dead creature.

When the DNA sequences from two different disease-causing versions of the human gene were compared with the sequence carried by people who do not have the disease, a consistent difference was found, as predicted by Garrod's interpretation. The difference between normal urine and alkaptonuria was as small as it could be: of the hundreds of amino acids that assemble in one of the enzymes used by the liver to break down a compound called tyrosine, one single amino acid was misplaced in each case. Remarkably, the amino acid that was different in patients with the disease was one that was the same in both healthy people and healthy bread molds. Such conserved features are likely to have been important for survival for a very long time. In the hundreds of millions of years since the age of the last common ancestor of humans and bread molds, copies of the gene must have accumulated many mutations, but each one that changed this amino acid must have so damaged the chances of survival that only species with the same amino acid have persisted to this day.

2. While it makes little sense to screen every woman's DNA for the presence of two competent BRCA1 or BRCA2 genes, once a family learns that some of its members carry a germ-line mutation for a gene like BRCA1, all the other members need to decide whether to submit their DNA to the same painless but fearful test. Also, people who share a common ancestry thereby share a particular ancestral selection of the many possible versions of each human gene in their germ lines. As a result, they will have a higher than usual probability of inheriting and passing on shared versions of genes. When a mutation in a gene like

BRCA1 or BRCA2 is present in one of the ancestral germ lines, everyone sharing that particular germ line thereafter will continue to have a higher than background risk of inheriting the mutation. For women from such high-risk groups, DNA diagnosis for these two genes — with all of its ambiguities, pitfalls, and tendencies to create problems rather than solve them — has become all but mandatory.

This situation has come to light with special poignancy for the Ashkenazim, the millions of people throughout the world who are the descendants of the Jews of the Pale, a region of Eastern and Central Europe in which they were obliged for centuries to live while neighboring lands excluded them. The incidence of BRCA1 and BRCA2 mutations is about one percent in Ashkenazim. A single dominant mutation in each gene is found in the vast majority of families of Ashkenazim tested, and that particular version of each gene is rarely found in people of other ethnic backgrounds. The inherited propensity to get breast cancer has led some people to call it a "Jewish disease," but as the chief rabbi of London once famously said in response to an article in the London *Times* about Tay-Sachs disease, "There are no Jewish diseases."

There cannot be, because although they claim a common ancestry, Jews are not in fact a single biological family; there are no DNA sequences common to all Jews and present only in Jews. Instead, the murderous intentions of strangers have fixed in the DNAs of Ashkenazim a history of their near extinction. So-called Jewish diseases like BRCA1/BRCA2 breast cancer and Tay-Sachs disease are evidence of a shared ancestry that has included periodic massacres of such ferocity that only a small number of families were able to survive. The utter sameness of mutations in BRCA1 and other genes in so many Ashkenazim suggests that the Jews whose ancestors came from the Pale — about nine of every ten Jews alive today — are the descendants of a small remnant of a few thousand families who survived a particularly devastating pogrom in the Pale of the mid-1600s.

As we are all susceptible to the mutations that bring on a cancer, one day everyone will be a candidate for a germ-line DNA test for cancer susceptibility, just as Ashkenazim are candidates for BRCA1 or BRCA2 testing today. But to what avail? The genetic differences that may lead to a better understanding of how to treat a tumor are simply not the same as the genetic differences between people that can be used only to tell someone his or her future. A blurring of this distinction is understandable as the wishful thinking of a frightened group of scientists unconsciously trying to keep cancer from striking their own bodies, but that does not make it right. The distinction needs to be made quite clear before it leads to great mischief: better DNA prognosis with neither explanation nor treatment is the worst of all possibilities. A DNA analysis of the versions of BRCA1 and BRCA2 genes in the general population has hardly any function at all, except to divide women into a minority, who will almost certainly get a breast tumor, and the rest, who have a one-in-nine chance of the same fate. Neither group can make much use of the information, since women in both groups still must undergo constant self-examination and since, in either group, detection of a tumor must be followed by the same harsh and painful treatment.

6. The Fear of Death

1. In an interview in the March 1997 *Scientific American*, Ron Graham, the chief scientist of AT&T, suggested a novel and quite simple *memento mori*, a quantitative way to see the coming of one's end. He takes a piece of ordinary graph paper and rules out a big box containing 100 boxes on each side. Every day he marks off one of the little boxes. Chances are, he says, that he will not live another thirty years; in that case, he will never finish filling in the big box on that sheet of graph paper.
2. Two of these conditions, Bloom and Werner syndromes, are both the consequence of inherited damage to proteins called

helicases, which open up a double-stranded stretch of chromosomal DNA to make copying, repair, and recombination easier; in bacteria and yeast cells, mutations in helicase generate chromosomes that are full of random errors.

3. Recent work to get around this problem has focused on synthetic variants of estrogen. With the 1997 discovery by the German scientist Christiane Behl that only a small portion of the estrogen molecule must be present to protect neuronal cells, we can imagine a wholly synthetic variant of estrogen able to stabilize the aging brain without causing any side effects. That would be an elegant, if indirect, form of "gene therapy" for aging, for it would work by sustaining the activity of a CAT gene that would otherwise fall silent in an old nerve cell, leading to its eventual death.

   An uncommon inherited susceptibility to Alzheimer's disease arises not from a version of a gene that leads directly to early neuron death but from a variant gene encoding one of the proteins that serve the brain, a member of a family of proteins called Apolipoprotein E, or ApoE. The ApoE proteins do not seem particularly relevant to brain function: they carry fat in the blood, and few of them are found in the brain and almost none in neurons. Nevertheless, people who inherit genes encoding a version of this protein called ApoE4 — but not the more common versions called ApoE2 and ApoE3 — have a high probability of developing Alzheimer's at an early age. Very few people who inherit two copies of the ApoE4 gene live out an average lifetime; most develop dementia well beforehand. The connection between ApoE4 and dementia is still circumstantial, but there are hints of a causal connection as well. A critical chemical difference between ApoE4 and the other versions of ApoE allows two molecules of ApoE4 to stick tightly to each other. These small clumps are fairly insoluble; they may play a role in seeding the larger clumps of insoluble protein that are found in the foci of dead neurons typical of Alzheimer's. No version of the ApoE gene completely protects a person from

eventually developing Alzheimer's disease: the incidence of dementia rises with advancing age, regardless of which ApoE versions are inherited.

4. Senescence in a dish may be a reflection of any normal cell's fate in a tissue or it may merely be an uninteresting consequence of continued stem-cell division outside the body in the absence of a full set of differentiating signals. The fibroblasts — stem cells for the differentiated tissue of a scar — must be retaining some chemical change indicating the passage of time, even after they have been removed from all of the body's hormonal signals and placed in the wholly unnatural environment of a dish. However, in the dish the stem cell produces a much larger number of descendants than it would in any tissue. When stem cells divide in a tissue, only one daughter persists as a stem cell while the other differentiates and dies. Even after fifty generations, the stem cells in each tissue will be no more numerous than when the tissue first reached adult proportions. Once taken from the tissue and put in a dish, the stem cell's descendants are freed from this constraint, and after each division both daughter cells retain the ability to divide again. After fifty such divisions, a single fibroblast in a dish will have generated enough cells able to divide again, to yield as many cells as there are in a person's entire body.

5. In our society, relatively free as it is from concerns about the propriety of manipulating the moment of death, the rights of the *gosess* would include at least one additional option. People who confuse life and death with rich and poor often argue that dying costs too much; among their favorite statistics is the true fact that about half the cost of hospital medical care is run up by people in the last six months of their lives. In a 1993 paper, the economics professor K. K. Tung argued — without irony — that as one way to reduce these costs, a dying person should be able to sell his right to treatment for a fee. Presumably the fee would be set at an amount less than the cost of living longer, but high enough to be an enticement to the survivors, if not the

dying person himself. Among the better responses to this article, one correspondent pointed out that if money were the issue — though it is, after all, not — then we should look to the one quarter of medical care costs that go today to administrative overhead. Since medical care costs in 1993 were about a million million dollars, that quarter fraction would be about $250 billion, rather more than it would cost to close the entire national debt in one year.

Further Reading

> I think that the man of science makes this mistake, and the mass of mankind along with him: that you should coolly give your chief attention to the phenomenon which excites you as something independent to you, and not as it is related to you. The important fact is its effect on me. . . .
>
> — Henry Thoreau, *Journal*, 5 November 1857

> Science is the criticism of myth. There would be no Darwin had there been no *Genesis*.
>
> — William Butler Yeats

Wide reading does not measure depth. As a wader can drown in a river only six inches deep on average, a reader of many topics in science can easily disappear into a sinkhole of details. It took me a lot of recent books and a few old ones — by Darwin, Freud, and Schrödinger, in particular — to see why it is difficult for the scientists who study the living world to accept certain implications of their own discoveries and for me to learn the details of these discoveries.

To help those of my own readers who may wish to study more deeply any of the topics I have covered, I have listed below some of the books that introduced me to each subject. Each reader is likely to turn to a different selection, but I am confident that everyone will be able to start with one of these.

For those readers who also wish to see my primary references, I have listed specialized books by chapter, along with a selection of the research papers that provided me with the material for my examples.

Books of General Interest

Appleyard, B., 1995. *Understanding the present: Science and the soul of modern man.* New York: Doubleday.

Bloom, H., 1989. *Ruin the sacred truths.* Cambridge, Mass.: Harvard University Press.

Darwin, C., 1859, reprinted 1985. *The origin of species by means of natural selection.* New York: Penguin.

Dronamraju, K., ed., 1995. *Haldane's Daedalus revisited.* New York: Oxford University Press.

Dubos, R., 1961. *The dreams of reason: Science and utopias.* New York: Columbia University Press.

———, 1970. *Reason awake: Science for man.* New York: Columbia University Press.

Freud, S., 1896. *Five lectures on psychoanalysis.* From *Standard Edition,* New York: Norton, 1961; paperback, 1987.

———, 1928. *The future of an illusion.* From *Standard Edition,* New York: Norton, 1961; paperback, 1987.

Gazzaniga, M., 1988. *Mind matters: How mind and brain interact to create our conscious lives.* Boston: Houghton Mifflin.

Ginzberg, E., 1990. *The medical triangle.* Cambridge, Mass.: Harvard University Press.

Goethe, W., 1801, 1832; reprinted 1991. *Faust.* New York: Penguin.

Horgan, J., 1996. *The end of science.* New York: Helix.

Jacob, F., 1973, reprinted 1982. *The logic of life.* New York: Pantheon.

———, 1973. *The statue within.* New York: Basic Books.

Johnson, G., 1995. *Fire in the mind: Science, faith and the search for order.* New York: Knopf.

Kandel, E., J. Schwartz, and T. Jessell, 1991. *Principles of neural science,* 3rd ed. New York: Appleton and Lange.

Kaufman, S., 1995. *At home in the universe: The search for the laws of self-organization and complexity*. New York: Oxford University Press.

Kitcher, P., 1992. *The advancement of science*. New York: Oxford University Press.

Lown, B., 1996. *The lost art of healing*. Boston: Houghton Mifflin.

McPhee, J., 1984. *Heirs of general practice*. New York: Farrar Straus Giroux.

Rolston, H., III, 1995. *Biology, ethics, and the origins of life*. Boston: Jones and Bartlett.

Sacks, O., 1995. *An anthropologist on Mars*. New York: Vintage.

Schrödinger, E., 1943, reprinted 1967. *What is life?: The physical aspects of the living cell*. Cambridge, Eng.: Cambridge University Press.

Soloveitchik, J., 1965, reprinted 1997. *The lonely man of faith*. New York: Aronson.

Solzhenitsyn, A., 1968. *The first circle*. New York: Bantam.

Weiner, J., 1995. *The beak of the finch*. New York: Vintage.

Chapter One

Books

Ackerman, D., 1990. *A natural history of the senses*. New York: Vintage.

Gould, S. J., 1987. *Time's arrow and time's cycle: Myth and metaphor in the discovery of geological time*. Cambridge, Mass.: Harvard University Press.

Hebb, D., 1949. *The origins of behavior*. New York: John Wiley.

Hubel, D. H., 1987. *Eye, brain, and vision*. New York: Scientific American Library.

Ladd-Franklin, C., 1929. *Colour and colour theories*. New York: Harcourt Brace.

Lamb, T., and J. Bourriau, 1995. *Colour: Art and science*. Cambridge, Eng.: Cambridge University Press.

Padgham, C., and J. Saunders, 1975. *The perception of light and colour*. New York: Academic.

Price, H., 1996. *Time's arrow and Archimedes' point: New directions for the physics of time.* New York: Oxford University Press.

Sacks, O., 1985. *The man who mistook his wife for a hat.* New York: Touchstone.

Tanner, J., 1990. *Fetus into man: Physical growth from conception to maturity.* Cambridge, Mass.: Harvard University Press.

Research articles

Buck, L., 1996. Information coding in the vertebrate olfactory system. *Ann Rev Neurosci* 19:517–44.

Churchland, P., and T. Sejnowski, 1988. Perspectives on cognitive neuroscience. *Science* 242:741.

Darlington, T., et al., 1998. Closing the circadian loop: CLOCK-induced transcription of its own inhibitors per and tim. *Science* 280:1599.

Dulac, C., et al., 1995. A novel family of genes encoding putative pheromone receptors in mammals. *Science* 83:195.

Edelman, G., 1976. Surface modulation in cell recognition and cell growth. *Science* 192:218.

Hall, B., 1995. Atavisms and atavistic mutations. *Nature Genetics* 10:126.

Hawking, S., and R. Penrose, 1996. The nature of space and time. *Scientific American* July:60.

Hunt, D., et al., 1995. The chemistry of John Dalton's color blindness. *Science* 265:984.

Ishai, A., et al., 1995. Common mechanisms of visual imagery and perception. *Science* 268:1772.

Matsunami, H., and L. Buck, 1997. A multigene family encoding a diverse array of putative pheromone receptors in mammals. *Cell* 90:775–84.

Maunsell, J., 1995. The brain's visible world: Representation of visual targets in the cerebral cortex. *Science* 270:764.

Milton, K., 1993. Diet and primate evolution. *Scientific American* August:86.

Nathans, J., et al., 1986. Molecular genetics of inherited variation in human color vision. *Science* 232:203–10.

Stern, K., et al., 1998. Regulation of ovulation by human pheromones. *Nature* 392:177.

Vanderhaeghen, P., et al., 1993. Olfactory receptors are displayed on dog mature sperm cells. *J Cell Biology* 123:1441.

Zeng, C., et al., 1996. A human axillary odorant is carried by apolipoprotein D. *Proc Nat Acad Sci* 93:6626.

Zhao, H., et al., 1998. Functional expression of a mammalian odorant receptor. *Science* 279:237.

Chapter Two

Books

Aertsen, A., ed., 1993. *Brain theory: Spatio-temporal aspects of brain function*. Amsterdam: Elsevier.

Blakemore, C., and S. Greenfield, 1987. *Mindwaves: Thoughts on intelligence, identity, and consciousness*. New York: Basil Blackwell.

Buszáki, G., et al., eds., 1994. *Temporal coding in the brain*. Berlin: Springer-Verlag.

Changeaux, J.-.P., et al., eds., *Origins of the human brain*. Oxford, Eng.: Clarendon Press.

Churchland, P., and T. Sejnowski, 1992. *The computational brain*. Cambridge, Mass.: MIT Press.

Cotterill, R. M. J., ed., 1989. *Models of brain function*. Cambridge, Eng.: Cambridge University Press.

Koch, C., and J. L. Davis, eds., 1994. *Large-scale neuronal theories of the brain*. Cambridge, Mass.: MIT Press.

Penrose, R., 1991. *The emperor's new mind: Concerning computers, minds, and the laws of physics*. New York: Penguin.

Young, R., 1990. *Mind, brain, and adaptation in the nineteenth century: Cerebral localization and its biological context from Gall to Ferrier*. New York: Oxford University Press.

RESEARCH ARTICLES

Alonso, J.-M., et al., 1996. Precisely correlated firing in cells of the lateral genticulate nucleus. *Nature* 382:815.

Antoch, M., et al., 1997. Functional identification of the mouse circadian *clock* gene by transgenic BAC rescue. *Cell* 89:655.

Barth, D., et al., 1996. Thalamic modulation of high-frequency oscillating potentials in auditory cortex. *Nature* 382:78.

Buszáki, G., 1991. The thalamic clock: Emergent network properties. *Neuroscience* 41:351.

Calvin, W., 1994. The emergence of intelligence. *Scientific American* October:101.

Cepko, C., 1992. What do progenitor cells tell their daughters during development of the cerebral cortex? *JNIH Research* 4:60.

De Koninck, P., and H. Schulman, 1998. Sensitivity of CaM Kinase II to the frequency of Ca2+ oscillatons. *Science* 279:227.

Di Prisco, G., 1984. Hebb synaptic plasticity. *Progress in Neuroscience* 22:89.

Dollins, A., et al., 1994. Effect of inducing nocturnal serum melatonin concentration in daytime on sleep, mood, body temperature and performance. *Proc Nat Acad Sci USA* 91:1824.

Gannon, P., et al., 1998. Asymmetry of chimpanzee plenum temporale: Humanlike pattern of Wernicke's brain language area homologue. *Science* 179:220.

Gerstner, W., et al., 1996. A neuronal learning role for sub-millisecond temporal coding. *Nature* 383:76.

Gray, C., et al., 1989. Oscillatory responses in cat visual cortex exhibit inter-columnar synchronization which reflects global stimulus properties. *Science* 338:334.

———, 1996. Chattering cells: Superficial pyramidal neurons contributing to the generation of synchronous oscillations in the visual cortex. *Science* 274:109.

Graziano, M., 1997. Coding the location of objects in the dark. *Science* 277:239.

Joliot, M., et al., 1994. Human oscillatory brain activity near 40 Hz coexists with cognitive temporal binding. *Proc Nat Acad Sci* 91:11748.

Just, M., et al., 1996. Brain activation modified by sentence comprehension. *Science* 274:114.

King, D., et al., 1997. Positional cloning of the mouse circadian *clock* gene. *Cell* 89:641.

Kopell, N., et al., 1994. Rhythmogenesis, amplitude modulation, and multiplexing in a cortical architecture. *Proc Nat Acad Sci* 91:10586.

Laurent, G., et al., 1994. Encoding of olfactory information with oscillating neural assemblies. *Science* 265:1872.

Libet, B., 1965. Cortical activation in conscious and unconscious experience. *Perspectives in Biology & Medicine* 9:77–86.

Libet, B., et al., 1983. Time of conscious intention to act in relation to onset of cerebral activity (readiness-potential). The unconscious initiation of a freely voluntary act. *Brain* 106:623.

———, 1991. Control of transition from sensory detection to sensory awareness in man by the duration of a thalamic stimulus. *Brain* 114:1731.

Llinás, R., et al., 1993. Coherent 40-Hz oscillation characterizes dream state in humans. *Proc Nat Acad Sci* 90:2078.

Pantev, C., et al., 1998. Increased auditory cortical representation in musicians. *Nature* 392:811.

Reppert, S., et al., 1994. Cloning and characterization of a mammalian melatonin receptor that mediates reproductive and circadian responses. *Cell* 13:1177.

Ribary, U., et al., 1991. Magnetic field tomography of coherent thalamocortical 40-Hz oscillations in humans. *Proc Nat Acad Sci* 88:11037.

Sack, R., 1998. Melatonin. *Science & Medicine* September 1998:8.

Saffran, J., et al., 1996. Statistical learning by 8-month-old infants. *Science* 274:1926.

Schechter, B., 1996. How the brain gets rhythm. *Science* 274:339.

Schwartz, W., 1996. Internal timekeeping. *Science & Medicine* May:44.

Sehgal, A., et al., 1995. Rhythmic expression of *timeless*: A basis for promoting circadian cycles in *period* gene autoregulation. *Science* 270:808.

Singer, W., and C. Gray, 1995. Visual feature integration and the temporal correlation hypothesis. *Ann Rev Neurosciences* 18:555.

Tosini, G., et al., 1996. Circadian rhythms in cultured mammalian brain. *Science* 272:419.

Traub, R., et al., 1996. A mechanism for generation of long-range synchronous fast oscillations in the cortex. *Nature* 382:621.

Ungerleiter, L., 1995. Functional brain imaging studies of cortical mechanisms for memory. *Science* 270:769.

van Turennout, M., et al., 1998. Brain activity during speaking: From syntax to phonology in 40 milliseconds. *Science* 280:572.

Yuste, R., et al., 1992. Neuronal domains in developing neo-cortex. *Science* 257:665.

Chapter Three

Books

American Psychiatric Association, 1994. *DSM-IV: Diagnostic and statistical manual of mental disorders*, 4th ed. Washington, D.C.: APA.

Arlow, J., 1969. *Unconscious fantasy and disturbances of conscious experience,* in *Psychoanalysis: Clinical theory and practice.*

Bickerton, D., 1995. *Language and human behavior*. Seattle: University of Washington Press.

Chalmers, D., 1996. *The conscious mind: In search of a fundamental theory*. New York: Oxford University Press.

Crick, F., 1994. *The astonishing hypothesis: The scientific search for the soul*. New York: Scribners.

Eccles, J., 1989. *Evolution of the brain*. London: Routledge.

Gay, P., *Freud: A life for our time*. New York: Anchor Books.

Gilman, S., 1988. *Disease and representation*. Ithaca: Cornell University Press.

Griffin, D., 1994. *Animal minds*. Chicago: University of Chicago Press.

Harrington, A., 1992. *So human a brain*. Boston: Birkhauser.

Jamison, K. R., 1995. *An unquiet mind: A memoir of moods and madness*. New York: Vintage.

Johnson, G., 1992. *In the palaces of memory*. New York: Vintage.

Kaplan, H., and B. Sadock, 1991. *Synopsis of psychiatry*, 6th ed. Baltimore: Williams and Wilkins.

Kitcher, Patricia, 1992. *Freud's dream: A complete interdisciplinary science of mind*. Cambridge, Mass.: MIT Press.

Klein, M., and J. Riviere, 1937, reprinted 1964. *Love, hate and reparation*. New York: Norton.

Kuper, A., 1994. *The chosen primate*. Cambridge, Mass.: Harvard University Press.

Luria, A., 1973. *The working brain*. New York: Basic Books.

Penrose, R., 1994. *Shadows of the mind: A search for the missing science of consciousness*. Oxford: Oxford University Press.

Plotkin, H., 1994. *Darwin machines and the nature of knowledge*. Cambridge, Mass.: Harvard University Press.

Ritvo, L., 1990. *Darwin's influence on Freud*. New Haven: Yale University Press.

Schaffer, R., 1976. *A new language for psychoanalysis*. New Haven: Yale University Press.

Solso, R., and D. Massaro, 1995. *The science of the mind*. New York: Oxford University Press.

Star, S. L., 1983. *Regions of the mind*. Palo Alto: Stanford University Press.

Sulloway, F., 1992. *Freud, biologist of the mind: Beyond the psychoanalytic legend*. Cambridge, Mass.: Harvard University Press.

Vaughan, S., 1997. *The talking cure: The science behind psychotherapy*. New York: Putnam.

Wolpert, L., 1994. *Unnatural nature of science*. Cambridge, Mass.: Harvard University Press.

Wright, R., 1994. *The moral animal: Evolutionary psychology and everyday life*. New York: Pantheon.

Research articles

Adolphs, R., et al., 1994. Impaired recognition of emotion in facial expressions following bilateral damage to the human amygdala. *Nature* 372:669.

Andreason, N., 1997. Linking mind and brain in the study of mental illnesses: A project for a scientific psychopathology. *Science* 275:1586.

Barinaga, C., 1992. The brain remaps its own contours. *Science* 258:216.

Castiello, U., et al., 1995. A brain-damaged patient with an unusual perceptuomotor deficit. *Nature* 374:805.

Chalmers, D., 1995. The puzzle of conscious experience. *Scientific American* December:80.

Crick, F., and C. Koch, 1992. The problem of consciousness. *Scientific American* 267:152.

Eichenbaum, H., 1997. How does the brain organize memories? *Science* 277:330.

Freud, S., 1901. On Dreams. In *Standard Edition*, New York: Norton, 1961.

———, 1907. Creative writers and daydreaming. In *Standard Edition*, New York: Norton, 1961.

———, 1907. Obsessional actions and religious practices. In *Standard Edition*, New York: Norton, 1961.

———, 1919. The "uncanny." In *Standard Edition*, New York: Norton, 1961.

Goodstein, W., 1992. Dreams of science. *Science* 258: 1503–4.

Greenwald, A., et al., 1996. Three cognitive markers of unconscious semantic activation. *Science* 273:1699.

Horgan, J., 1996. Why Freud isn't dead. *Scientific American* 172:106.

Isaacs, S., The nature and function of phantasy. In *Developments in Psychoanalysis*, ed. J. Riviere. 82–98.

Kandel, E., 1979. Psychotherapy and the single synapse. *New England J Med* 301:1028.

Karni, A., 1997. Adult cortical plasticity and reorganization. *Science & Medicine* 1/97:24.

Knowlton, B., et al., 1996. A neostriatal habit learning system in humans. *Science* 273:1399.

Kornberg, A., 1992. Science is great but scientists are still people. *Science* 257:859.

Kully, J., and C. Koch, 1991. Does anesthesia cause loss of consciousness? *Trends in Neurological Sciences* 14:6.

LeDoux, J., 1994. Emotion, memory and the brain. *Scientific American* June:50.

McHugh, P., 1994. Psychiatric misadventures. *American Scholar* 61:497.

Michels, A., 1995. From transference to metaphor. *Clinical Studies: International Journal of Psychoanalysis* 1:1–15.

Olds, D., 1992. Consciousness: A brain-centered approach. *Psychoanalytic Inquiry* 12:419–44.

———, 1993. Connectionism and psychoanalysis. *J Amer Psychoanalytical Assn* 42:581–611.

Sapolsky, R., 1994. Measures of life. *The Sciences* 3/94:10.

Schultz, W., et al., 1997. A neural substrate of prediction and reward. *Science* 275:1997.

Schwartz, J., 1997. Obsessive-compulsive disorder. *Science & Medicine* March:14.

Searle, J., 1990. Consciousness, explanatory inversion and cognitive science. *Behavioral and Brain Sciences* 13:585–642.

Seidenberg, M., 1997. Language acquisition and use: Learning and applying probabilistic constraints. *Science* 275:1599.

Seyfarth, R., et al., 1992. Meaning and mind in monkeys. *Scientific American* December:122–26.

Solms, M., 1995. Is the brain more real than the mind? *Psychoanl Psychother* 9:107.

———, 1995. *Six lectures on Freud, Luria, neuranatomy and brain function.* Sponsored by the New York Psychoanalytic Association and the British Psychoanalytic Association.

Terrace, H., 1985. In the beginning was the "name." *Amer Psychologist* 40:1011–28.

Velmans, M., 1991. Is human information processing conscious? *Behavioral and Brain Sciences* 14:651.

Chapter Four

Books

Biddle, W., 1996. *A field guide to the germs*. New York: Anchor Books.

Bloom, B., ed., 1994. *Tuberculosis: Pathogenesis, protection, and control*. Washington, D.C.: ASM Press.

Despommier, D., et al., eds., 1995. *Parasitic diseases*, 3rd ed. New York: Springer.

Dixon, B., 1994. *Power unseen: How microbes rule the world*. San Francisco: Freeman.

———. International Conference on Emerging Infectious Diseases, 1998. Proceedings. *Emerging infectious diseases* 4:Issue 3.

Evans, A., ed., 1991. *Viral infections of humans: Epidemiology and control*. New York: Plenum.

Evans, A., et al., eds., 1991. *Bacterial infections of humans: Epidemiology and control*. New York: Plenum.

Hope-Simpson, E., 1992. *The transmission of epidemic influenza*. New York: Plenum.

Lederberg, J., et al., 1992. *Emerging infections: Microbial threats to health in the United States*. Washington, D.C.: National Academy Press.

Margulis, L., 1983, reprinted 1991. *Symbiosis in cell evolution: Microbial communities in the Archean and Proterozoic eons*. New York: Freeman.

———, 1984. *Early life*. Boston: Jones and Bartlett.

Morse, S., ed., 1993. *Emerging viruses*. New York: Oxford University Press.

Postgate, J., 1994. *The outer reaches of life*. New York: Cambridge University Press.

Smith, J., and E. Szathmary, 1995. *The major transitions in evolution*. New York: Freeman.

Sober, E., ed., 1994. *Conceptual issues in evolutionary biology*, 2nd ed. Cambridge, Mass.: MIT Press.

Williams, G., 1991. *Natural selection: Domains, levels and challenges*. New York: Oxford University Press.

Wilson, M., et al., eds., 1994. Disease in evolution: Global changes and emergence of infectious diseases. *Annals of the New York Academy of Sciences* 740. New York: NYAS.

Research articles

Balaban, N., et al., 1998. Autoinducer of virulence as a target for vaccine and therapy against Staphylococcus aureus. *Science* 280:438–40.

Banerjee, A., et al., 1994. *inhA*, a gene encoding a target for isoniazid and ethionamide in Mycobacterium tuberculosis. Science 263:27.

Beardsley, T., 1995. Better than a cure. *Scientific American* January:88.

Behbehani, A., 1991. The smallpox story: Historical perspective. *Am Soc Microbiol News* 57:571.

Belisle, J., et al., 1997. Role of the major antigen of *Mycobacterium tuberculosis* in cell wall biogenesis. *Science* 276:1420.

Berkelman, R., et al., 1994. Infectious disease surveillance: A crumbling foundation. *Science* 264:368.

Bieber, D., et al., 1998. Type IV pili, transient bacterial aggregates, and virulence of enteropathic Escherichia coli. *Science* 280:2114.

Black, P., 1995. Psychoneuroimmunology: Brain and immunity. *Science & Medicine* November:16.

Bloom, B., 1994. The United States needs a national vaccine authority. *Science* 265:1378.

Blower, S., et al., 1994. Prophylactic vaccines, risk behavior change, and the probability of eradicating HIV from San Francisco. *Science* 265:1451.

Boren, T., et al., 1993. Attachment of *Helicobacter pylori* to human gastric epithelium mediated by blood group antigens. *Science* 262:1892.

Breiman, R., et al., 1994. Emergence of drug-resistant pneumococcal infections in the United States. *J Amer Med Assn* 271:1831.

Bry, L., et al., 1996. A model of host-microbial interactions in an open mammalian ecosystem. *Science* 273:1380.

Chen, L.-M., et al., 1996. Requirement of CDC42 for *Salmonella*-induced cytoskeletal and nuclear responses. *Science* 274:2115.

Colwell, R., 1996. Global climate and infectious disease: The global paradigm. *Science* 274:2025.

Cookson, W., et al., 1997. Asthma: An epidemic in the absence of infection? *Science* 275:41.

Dean, M., et al., 1996. Genetic restriction of HIV-1 infection and progression to AIDS by a deletion allele of the *CKR5* structural gene. *Science* 273:1856.

Dubois, A., 1995. Spiral bacteria in the human stomach: The gastric helicobacters. *Emerging Infectious Diseases* 1:79.

Eigen, M., 1993. Viral quasispecies. *Scientific American* July:42.

Evans, D., et al., 1994. Economics and the argument for parasitic disease control. *Science* 265:1868.

Ewald, P., 1996. Guarding against the most dangerous emerging pathogens: Insights from evolutionary biology. *Emerging Infectious Diseases* 2:245.

Farmer, P., 1996. Social inequalities and emerging infectious diseases. *Emerging Infectious Diseases* 2:259.

Fidler, D., 1996. Globalization, international law, and emerging infectious diseases. *Emerging Infectious Diseases* 2:77.

Finlay, B., and P. Cossart, 1997. Exploitation of mammalian host cell functions by bacterial pathogens. *Science* 276:718.

Finlay, B., et al., 1989. Epithelial cell surfaces induce *Salmonella* proteins required for bacterial adherence and invasion. *Science* 243:940.

Fritz, C., et al., 1996. Surveillance for pneumonic plague in the United States during an international emergency: A model for control of imported emerging diseases. *Emerging Infectious Diseases* 2:30.

Goebel, S., et al., 1990. The complete DNA sequence of Vaccinia virus. *Virology* 179:247.

Goudsmit, J., et al., 1991. Genomic diversity and antigenic variation of HIV-1: Links between pathogenesis, epidemiology and vaccine development. *FASEB J* 5:2427.

Hilleman, M., 1969. Toward control of viral infections of man. *Science* 164:506.

Janeway, C., 1993. How the immune system recognizes invaders. *Scientific American* September:73.

Joklik, W., et al., 1993. Why the smallpox virus stocks should not be destroyed. *Science* 262:1225.

Krogstad, D., et al., 1987. Efflux of cholorquine from Plasmodium falciparum: Mechanism of chloroquine resistance. *Science* 238:1283.

Lederberg, J., 1988. Medical science, infectious disease, and the unity of humankind. *J Amer Med Assn* 260:684.

———, 1988. Pandemic as a natural evolutionary phenomenon. *Social Research* 55:343.

Le Guenno, B., 1995. Emerging viruses. *Scientific American* October:56.

Leutwyler, K., 1995. The price of prevention. *Scientific American* April:124.

Losick, R., and D. Kaiser, 1997. How and why bacteria communicate. *Scientific American* February:68.

Mahy, B., et al., 1993. The remaining stocks of smallpox should be destroyed. *Science* 262:1223.

Marrack, P., et al., 1993. How the immune system recognizes the body. *Scientific American* September:81.

McMeekin, T., et al., 1997. Quantitative microbiology: A basis for food safety. *Emerging Infectious Diseases* 3:541.

Melnick, J., et al., 1993. Cytomegalovirus and atherosclerosis. *European Heart J* 14:Supplement 30.

Miller, J., et al., 1989. Coordinate regulation and sensory transduction in the control of bacterial virulence. *Science* 243:916.

Miller-Hjelle, M., et al., 1997. Polycystic kidney disease: An unrecognized emerging infectious disease? *Emerging Infectious Diseases* 3:113.

Neu, H., 1992. The crisis in antibiotic resistance. *Science* 257:1064.

Nichol, S., et al., 1993. Genetic identification of a Hantavirus associated with an outbreak of acute respiratory distress. *Science* 262:914–17.

Niklasson, B., et al., 1998. Could myocarditis, insulin-dependent diabetes mellitus, and Guillain-Barré syndrome be caused by one or more infectious agents carried by rodents? *Emerging Infectious Diseases* 4:187.

Nowack, M., et al., 1995. How HIV defeats the immune system. *Scientific American* August:58.

Nussenzweig, R., et al., 1994. Malaria vaccines: Multiple targets. *Science* 265:1381.

O'Callaghan, E., et al., 1991. Schizophrenia after prenatal exposure to 1957 A2 influenza epidemic. *Lancet* 337:1250.

Orenstein, J., et al., 1997. Macrophages as a source of HIV during opportunistic infections. *Science* 276:1857.

Pace, N., 1997. A molecular view of microbial diversity and the biosphere. *Science* 276:734.

Pascopella, L., et al., 1994. Use of in vivo complementation in *Mycobacterium tuberculosis* to identify a genomic fragment associated with virulence. *Infection and Immunity* 62:1313.

Paul, W., 1993. Infectious diseases and the immune system. *Scientific American* September:91.

Pettersson, J., et al., 1996. Modulation of Virulence Factor expression by pathogen target cell contact. *Science* 273:1231.

Saikku, P., et al., 1992. Chronic *Chlamydia pneumoniae* infection as a risk factor for coronary heart disease in the Helsinki Heart Study. *Annals Internal Med* 116:273.

Satcher, D., 1995. Emerging infections: Getting ahead of the curve. *Emerging Infectious Diseases* 1:1.

Shlaes, D., et al., 1991. Antimicrobial resistance: New directions. *Am Soc Microbiol News* 57:455.

Slater, A., and A. Cerami, 1992. Inhibition by chloroquine of a novel haem polymerase enzyme activity in malaria trophozoites. *Science* 355:167.

Snider, D., et al., 1994. Multi-drug-resistant tuberculosis. *Science & Medicine* May:16.

Sprent, J., et al., 1994. Lymphocyte life-span and memory. *Science* 265:1395.

Stover, C., et al., 1991. New use of BCG for recombinant vaccines. *Nature* 351:456.

Streit, W., et al., 1995. The brain's immune system. *Scientific American* November:54.

Tacket, C., 1998. Immunogenicity in humans of a recombinant bacterial antigen delivered in a transgenic potato. *Nature Medicine* 4:607.

Taubenberger, J., et al., 1997. Initial genetic characterization of the 1918 "Spanish" influenza virus. *Science* 275:1793.

Valero, M., et al., 1993. Vaccination with SPf66, a chemically synthesised vaccine, against *Plasmodium falciparum* malaria in Colombia. *Lancet* 341:705.

Wedemeyer, G., et al., 1997. Structural insights into the evolution of an antibody combining site. *Science* 276:1665.

Wigzell, H., 1993. The immune system as a therapeutic agent. *Scientific American* September:127.

Wilson, M., 1995. Travel and the emergence of infectious diseases. *Emerging Infectious Diseases* 1:39.

Zheng, L., et al., 1991. Low-resolution genome map of the malaria mosquito *Anopheles gambiae*. *Proc Nat Acad Sci* 88:11187.

Zhu, T., et al., 1998. An African HIV-1 sequence from 1959 and implications for the origin of the epidemic. *Nature* 391:594.

Chapter Five

Books

Davis, B., 1991. *The genetic revolution: Scientific prospects and public perceptions*. Baltimore: Johns Hopkins University Press.

Harden, V., 1986. *Inventing the NIH: Federal biomedical research policy 1887–1937*. Baltimore: Johns Hopkins University Press.

Kitcher, P., 1995. *The lives to come*. New York: Simon and Schuster.

Nelkin, D., and L. Tancredi, 1989. *Dangerous diagnostics: The social power of biological information*. New York: Basic Books.

Nesse, R., and G. Williams, 1994. *Why we get sick: The new science of Darwinian medicine*. New York: Times Books.

Pollack, R., 1994. *Signs of life: The language and meanings of DNA*. Boston: Houghton Mifflin.

Research articles

Aaltonen, L., et al., 1993. Clues to the pathogenesis of familial colorectal cancer. *Science* 260:812–16.

Adams, M., et al., 1995. Initial assessment of human gene diversity and expression patterns based upon 83 million nucleotides of cDNA sequence. *Nature* 337:supplement page 3.

Aguilar, F., et al., 1994. Geographic variation of *p53* mutational profile in nonmalignant human liver. *Science* 264:1317.

Bailar, J., et al., 1997. Cancer undefeated. *New England J Med* 336: 1569; response from B. Kramer and R. Klausner, 337: 931.

Benjamin, J., et al., 1996. Population and familial association between the D4 dopamine receptor gene and measures of Novelty Seeking. *Nature genetics* 12:81.

Bianchi, D., et al., 1996. Male fetal progenitor cells persist in maternal blood for as long as 27 years postpartum. *Proc Nat Acad Sci* 93:705.

Billings, P., et al., 1992. Discrimination as a consequence of genetic testing. *American J Human Genetics* 50:476.

Bouchard, T., 1994. Genes, environment and personality. *Science* 264:1700.

Brinster, R., et al., 1994. Germline transmission of donor haplotype following spermatogonial transplantation. *Proc Nat Acad Sci* 91:11303.

Bronner, E., et al., 1994. Mutation in the DNA mismatch repair gene homologue hMLH1 is associated with Hereditary non-polyposis colon cancer. *Nature* 368:258–61.

Bryan, T., et al., 1995. Telomere elongation in immortal cells without detectable telomerase activity. *Embo J* 14:4240.

Buetow, K., et al., 1994. Integrated human genome-wide maps constructed using the CEPH reference panel. *Nature Genetics* 6:391–93.

Cairns, J., 1985. The treatment of diseases and the war against cancer. *Scientific American* November:51.

Capecchi, M., 1994. Targeted gene replacement. *Scientific American* March:52.

Cheung, M.-C., et al., 1996. Prenatal diagnosis of sickle cell anaemia and thalassaemia by analysis of fetal cells in maternal blood. *Nature Genetics* 14:264.

Cohn, D., et al., 1986. Lethal osteogenesis imperfecta resulting from a single nucleotide change in one human pro-a1(I) collagen allele. *Proc Nat Acad Sci* 83:6045–47.

Cole, P., et al., 1996. Declining cancer mortality in the United States. *Cancer* 78:2045.

Davies, H., et al., 1996. The effect of human serum paraoxonase is reversed with diazoxon, soman and sarin. *Nature Genetics* 14:334.

Davis, D., et al., 1997. Environmental influences on breast cancer risk. *Science & Medicine* May:56.

Deng, G., et al., 1996. Loss of heterozygosity in normal tissue adjacent to breast carcinomas. *Science* 274:2057.

Denissenko, M., et al., 1996. Preferential formation of benzo[alpha]pyrene adducts at lung cancer mutational hotspots. *Science* 274:430.

Donehower, L., et al., 1992. Mice deficient for p53 are developmentally normal but susceptible to spontaneous tumours. *Nature* 356:215.

Duesbery, P., et al., 1998. Proteolytic inactivation of MAP-kinase-kinase by anthrax lethal factor. *Science* 280:734.

Duke, R., et al., 1996. Cell suicide in health and disease. *Scientific American* December:80.

Ebstein, R., et al., 1996. Dopamine D4 receptor (D4DR) exon III polymorphism associated with the human personality trait of Novelty Seeking. *Nature Genetics* 12:78.

Feng, J., et al., 1995. The RNA component of human telomerase. *Science* 269:1236.

Fernandez-Cañon, J., et al., 1996. The molecular basis of alkaptonuria. *Nature Genetics* 14:1924.

Fodor, S., 1997. Massively parallel genomics. *Science* 277:393.

Freimer, N., et al., 1996. Genetic mapping using haplotype, association and linkage methods suggests a locus for severe bipolar disorder (BPI) at 18q22-q23. *Nature Genetics* 12:436.

Garrod, A., 1902. The incidence of alkaptonuria: A study in clinical individuality. *Lancet* 2: 1616–20.

Giardiello, F., et al., 1997. The use and interpretation of commercial APC testing for familial adenomatous polyposis. *New England J Med* 336:823.

Gupta, J., et al., 1996. Development of retinoblastoma in the absence of telomerase activity. *J National Cancer Inst* 88:1152.

Ionov, Y., et al., 1993. Ubiquitous somatic mutations in simple repeated sequences reveal a new mechanism for colonic carcinogenesis. *Nature* 363:558–61.

Kemp, C., et al., 1994. p53-deficient mice are extremely susceptible to radiation-induced tumorigenesis. *Nature Genetics* 8:66.

Kiesewetter, S., et al., 1993. A mutation in CFTR produces different phenotypes depending on chromosomal background. *Nature Genetics* 5:274.

Kim, N., et al., 1994. Specific association of human telomerase activity with immortal cells and cancer. *Science* 266:2011.

Leffell, D., et al., 1996. Sunlight and skin cancer. *Scientific American* July:52.

Lesch, K.-P., et al., 1996. Association of anxiety-related traits with a polymorphism in the serotonin transporter gene regulatory region. *Science* 274:1527.

Lingner, J., et al., 1997. Reverse transcriptase motifs in the catalytic subunit of telomerase. *Science* 276:561.

Malkin, D., et al., 1990. Germ line p53 mutations in a familial syndrome of breast cancer, sarcomas and other neoplasms. *Science* 250:1233.

Marcand, S., et al., 1997. A protein-counting mechanism for telomere length in yeast. *Science* 275:986.

Marcel, T., and J. Grausz, 1997. The TMC worldwide gene therapy enrollment report, end 1996. *Human Gene Therapy* 8:775.

Miki, Y., et al., 1994. A strong candidate for the breast and ovarian cancer susceptibility gene BRCA1. *Science* 266:66–71.

Perera, F., 1996. Uncovering new clues to cancer risk. *Scientific American* May:54.

Prives, C., 1994. How loops, beta sheets and alpha helices help us to understand p53. *Cell* 78:543.

Reiss, A., et al., 1995. Contribution of the FMR1 gene mutation to human intellectual dysfunction. *Nature Genetics* 11:331.

Risch, N., 1990. Genetic linkage and complex diseases, with special reference to psychiatric disorders. *Genetic Epidemiology* 7:3.

Risch, N., et al., 1995. Genetic analysis of idiopathic torsion dystonia in Ashkenazi Jews and their recent descent from a small founder population. *Nature Genetics* 9:152.

Roa, B., et al., 1996. Ashkenazi Jewish population frequencies for common mutations in BRCA1 and BRCA2. *Nature Genetics* 14:185.

Savitsky, K., et al., 1995. A single Ataxia Telangectasia gene with a product similar to PI-3 kinase. *Science* 268:1749.

Sidransky, D., et al., 1994. Clonal expansion of p53 mutant cells is associated with tumor progression. *Nature* 355:846.

Stillman, B., 1996. Cell cycle control of DNA replication. *Science* 274:1659.

Struewing, P., et al., 1995. The carrier frequency of the *BRCA1* 185delAG mutation is approximately 1 percent in Ashkenazi Jewish individuals. *Nature Genetics* 11:198.

Umar, A., et al., 1994. Defective mismatch repair in extracts of colorectal and endometrial cancer cell lines exhibiting microsatellite instability. *J Biol Chem* May 20:14367–70.

van Wezel, T., et al., 1996. Gene interaction and single gene effects in colon tumour susceptibility in mice. *Nature Genetics* 14:468.

Wang, D., et al., 1998. Large-scale identification, mapping and genotyping of single-nucleotide polymorphisms in the human genome. *Science* 280:1077.

Weber, B., 1996. Genetic testing for breast cancer. *Science & Medicine* January:12.

Wivel, N., et al., 1993. Germ-line gene modification and disease prevention: Some medical and ethical perspectives. *Science* 262:533.

Ziegler, A., et al., 1994. Sunburn and p53 in the onset of skin cancer. *Nature* 372:773.

Chapter Six

Books

Becker, E., 1975. *The denial of death*. New York: Free Press.

Bowker, J., 1993. *The meanings of death*. Cambridge, Eng.: Canto.

Callahan, D., 1987. *Setting limits*. New York: Simon and Schuster.

Cart, K., et al., 1992. *Human aging and chronic disease*. Boston: Jones and Bartlett.

Cohen, J., 1995. *How many people can the earth support?* New York: Norton.

Dalley, S., trans., 1984. *Myths from Mesopotamia: Creation, the flood, Gilgamesh and others*. New York: Oxford University Press.

Frankl, V., 1984. *Man's search for meaning*. New York: Washington Square Press.

Gillman, N., 1997. *The death of death*. Woodstock, Vt.: Jewish Lights.

Grollman, E., ed., 1974. *Concerning death: A practical guide for the living*. Boston: Beacon Press.

Kanungo, M., 1994. *Genes and aging*. Cambridge, Eng.: Cambridge University Press.

Keller, E. F., 1992. *Secrets of life, secrets of death*. New York: Routledge.

Kierkegaard, S., reprinted 1953. *Fear and trembling* and *Sickness unto death*. New York: Anchor.

Kübler-Ross, E., 1969, reprinted 1993. *On death and dying*. New York: Collier Macmillan.

Livi-Bacci, M., 1992. *A concise history of world populations*. Trans. New York: Blackwell.

Moore, T., 1993. *Lifespan: Who lives longer and why*. New York: Simon and Schuster.

Riley, J., 1990. *Sickness, recovery and death*. Iowa City: University of Iowa Press.

Rose, M., 1993. *Evolutionary biology of aging*. New York: Oxford University Press.

Simpopoulos, A., et al., 1993. *Genetic nutrition*. New York: Macmillan.

Waggoner, P., 1994. *How much land can ten billion people spare for nature?* Ames, Iowa: Council for Agricultural Science and Technology.

Research articles

Alzheimer's Disease Collaborative Group, 1995. The structure of the presenilin 1 (S182) gene and identification of six novel mutations in early onset AD families. *Nature Genetics* 11:219.

Bartus, R., et al., 1978. Aging in the Rhesus monkey: Debilitating effects on short-term memory. *J Gerontol* 33:858.

Behl, C., et al., 1997. Neuroprotection against oxidative stress by estrogens: Structure-activity relationship. *Molecular Pharmacology* 51:535.

Berardi, A., et al., 1995. Functional isolation and characterization of human hematopoietic stem cells. *Science* 267:104.

Bongaarts, J., 1994. Can the growing human population feed itself? *Scientific American* March:36.

Brown, J., and E. Pollitt, 1996. Malnutrition, poverty and intellectual development. *Scientific American* February:38.

Chiu, C.-P., et al., 1997. Replicative senescence and cell immortality: The role of telomeres and telomerase. *Proc Soc Exp Biol and Med* 214:99.

Curtsinger, J., 1995. Genetic variation and aging. *Ann Rev Genetics* 29:553.

Davis, R., et al., 1997. Mutations in mitochondrial cytochrome *c* oxidase genes segregate with late-onset Alzheimer disease. *Proc Nat Acad Sci* 94:4526.

Feeney, G., 1994. Fertility decline in East Asia. *Science* 266:518.

Freis, J., 1980. Aging, natural death, and the compression of mortality. *New England J Med* 303:130.

Fung, K., 1995. Dying for money: Overcoming moral hazard in

terminal illnesses through compensated physician-assisted death. *J Economics and Sociology* 52:275.

Gibbons, G., et al., 1996. Molecular therapies for vascular diseases. *Science* 272:689.

Gibbs, W., 1997. Profile: Ronald L. Graham. *Scientific American* March:28.

Harrington, L., et al., 1997. A mammalian telomerase-associated protein. *Science* 275:973.

Hart, R., and W. Setlow, 1974. Correlation between DNA excision-repair and life-span in a number of mammalian species. *Proc Nat Acad Sci* 71:2169.

Hastie, N., et al., 1990. Telomere reduction in human colorectal carcinoma and with aging. *Nature* 346:866.

Holliday, R., 1988. Toward a biological understanding of the aging process. *Perspectives in Biology and Medicine* 32:109.

Horiuchi, S., 1992. Stagnation in the decline of world population growth rate during the 1980s. *Science* 257:761.

Jazwinski, M., 1997. Longevity, genes and aging. *Science* 273:54.

Jenuth, J., et al., 1996. Random genetic drift in the female germline explains the rapid segregation of mammalian mitochondrial DNA. *Nature Genetics* 14:146–51.

Jitsukawa, M., et al., 1994. Birth control in Japan: Realities and prognosis. *Science* 265:1048.

Knaus, W., et al., 1991. Short-term mortality predictions for critically-hospitalized adults: Science and ethics. *Science* 254:389.

LaFerla, F., et al., 1995. The Alzheimer's A-beta peptide induces neurodegeneration and apoptotic cell death in transgenic mice. *Nature Genetics* 19:21.

Lithgow, G., et al., 1997. Mechanisms and evolution of aging. *Science* 273:80.

Makita, Z., et al., 1992. Hemoglobin-AGE: A circulating marker of advanced glycosylation. *Science* 258:651.

Marcand, S., et al., 1997. A protein-counting mechanism for telomere length regulation in yeast. *Science* 275:986.

Miller, R., 1997. The aging immune system: Primer and prospectus. *Science* 273:70.

Miyata, M., et al., 1996. Apolipoprotein E allele-specific antioxidant activity and effects on cytotoxicity by oxidative insults and beta-amyloid peptides. *Nature Genetics* 14:55.

Moraes, C., et al., 1993. A mitochondrial tRNA anticodon swap associated with a muscle disease. *Nature Genetics* 4:284–88.

Motulsky, A., 1987. Human genetic variation and nutrition. *American J Clinical Nutrition* 45:1108.

Olshansky, J., et al., 1990. In search of Methuselah: Estimating the upper limits to human longevity. *Science* 250:634.

———, 1993. The aging of the human species. *Scientific American* September:46.

Piel, G., 1994. AIDS and population "control." *Scientific American* February:124.

Prockop, D., 1997. Marrow stromal cells as stem cells for nonhematopoietic tissues. *Science* 276:71.

Raff, M., et al., 1993. Programmed cell death and the control of cell survival: Lessons from the nervous system. *Science* 262:695.

Robey, B., et al., 1993. The fertility decline in developing countries. *Scientific American* December:60.

Rosa, L., et al., 1998. A close look at therapeutic touch. *J Amer Med Assn* 279:1005.

Roses, A., 1995. Apolipoprotein E and Alzheimer disease. *Science & Medicine* September:16.

Rusting, R., 1992. Why do we age? *Scientific American* December:130.

Schächter, F., et al., 1994. Genetic associations with human longevity at the *APOE* and *ACE* loci. *Nature Genetics* 6:29.

Sohal, R., et al., 1997. Oxidative stress, caloric restriction, and aging. *Science* 273:59.

Sinclair, D., et al., 1997. Accelerated aging and nucleolar fragmentation in yeast sgs1 mutants. *Science* 277:1313.

Solomon, M., et al., 1993. Decisions near the end of life: Professional views on life-sustaining treatments. *American J Public Health* 33:14.

Szilard, L., 1959. On the nature of the aging process. *Proc Nat Acad Sci* 45:30.

ter Muelen, R., et al., 1994. What do we owe the elderly? *Hastings Center Report* 24:S1.

Tsevat, J., et al., 1998. Health values of hospitalized patients 80 years or older. *J American Medical Assn* 279:371.

Vaupel, J., 1998. Biodemographic trajectories of longevity. *Science* 280:855.

van Leeuwen, F., et al., 1998. Frameshift mutants of beta-amyloid precursor protein and ubiquitin-B in Alzheimer's and Down patients. *Science* 279:242.

van Steensel, B., and T. de Lange, 1997. Control of telomere length by the human telomeric protein TRF1. *Nature* 285:740.

Vita, A., et al., 1998. Aging, health risks, and cumulative mortality. *New England J Med* 338:1035.

Wallace, D., 1997. Mitochondrial DNA and disease. *Scientific American* August:40.

Wei, G., 1994. The germline: Familiar and newly uncovered properties. *Annual Reviews Genetics* 28:309.

Weindruch, R., 1996. Caloric restriction and aging. *Scientific American* January:46.

White, K., et al., 1996. How many Americans are alive because of twentieth-century improvements in mortality? *Population and Development Review* 22:415.

Willet, W., 1994. Diet and health: What should we eat? *Science* 264:532.

Wilmut, I., et al., 1997. Viable offspring derived from fetal and adult mammalian cells. *Nature* 385:810.

Wise, P., et al., 1997. Menopause: Aging of multiple pacemakers. *Science* 273:7069.

Youdim, Y., et al., 1997. Understanding Parkinson's Disease. *Scientific American* January:52.

Yu, C.-E., et al., 1996. Positional cloning of the Werner's syndrome gene. *Science* 272:258.

Conclusion

Duyao, M., et al., 1993. Trinucleotide repeat length instability and age of onset in Huntington's disease. *Nature Genetics* 4:387–92.

Fu, Y., et al., 1992. An unstable triple repeat in a gene related to myotonic muscular dystrophy. *Science* 255:1256–58.

Hudson, K., et al., 1995. Genetic discrimination and health insurance: An urgent need for reform. *Science* 270:391.

Müller-Hill, B., 1993. Science, truth and other values. *Quarterly Rev Biol* 68:399.

Pardes, H., et al., 1999. Effects of medical research on health care and the economy. *Science* 283:36.

Solomon, M., et al., 1991. Toward an expanded vision of clinical ethics education: From the individual to the institution. *Kennedy Institute of Ethics J* 1:225.

Appendix

Abelson, P., 1993. Policies for science and technology. *Science* 260:735.

Aggleton, P., et al., 1994. Risking everything? Risk behavior, behavior change, and AIDS. *Science* 265:341.

Andino, P., et al., 1994. Engineering poliovirus as a vaccine vector for the expression of diverse antigens. *Science* 265:1448.

Arthur, W. B., 1997. How fast is technology evolving? *Scientific American* February:105.

Cohen, J., et al., 1996. The medical expenditure panel survey: A national health information resource. *Inquiry* 33:373.

Cohen, L., et al., 1994. Privatizing public research. *Scientific American* September:72.

Collins, F., 1997. Family history and genetic risk factors: Forward to the future. *J Amer Med Assn* 278:1284.

Council for Responsible Genetics, 1993. Position paper on human germ line manipulation. Human Genetics Committee, Fall, 1992. *Human Gene Therapy* 4:35.

Editorial, 1994. Crossing the divide from vaccine technology to vaccine delivery. *J Amer Med Assn* 272:1138.

Editorial, 1997. Cancer wars on all fronts. *Nature Genetics* 15:221.

Godal, T., 1994. Fighting the parasites of poverty: Public research, private industry, and tropical diseases. *Science* 264:1864.

Hamerman, D., et al., 1992. Responses of the health professions to the demographic revolution: A multidisciplinary perspective. *Perspectives in Biology and Medicine* 35:583.

Harrison, P., and J. Lederberg, 1997. *Orphans and incentives: Developing technologies to address emerging and re-emerging infections*. Washington, D.C.: National Academy Press.

Haynes, B., 1993. Scientific and social issues of HIV vaccine development. *Science* 260:1279.

Kaplitt, M., et al., 1994. Long-term gene expression and phenotypic correction using adeno-associated virus vectors in the mammalian brain. *Nature Genetics* 8:148.

Kurland, C., 1997. Beating scientists into plowshares. *Science* 276:761.

Lander, E., 1996. The new genomics: Global views of biology. *Science* 274:536.

Lapham, E., et al., 1996. Genetic discrimination: Perspectives of consumers. *Science* 274:621.

Mason, H., et al., 1995. Expression of Hepatitis B surface antigen in transgenic plants. *Proc Nat Acad Sci* 89:11745.

Miller, T., et al.,1994. Medical-care spending — United States. *Morbidity and Mortality Weekly Reports* 43:581.

Mitcheson, A., 1993. Will we survive? *Scientific American* September:136.

Murray, C., et al., 1998. *U.S. patterns of mortality by county and by race, 1965–1994*. Cambridge, Mass.: Harvard School of Public Health.

NIAID international centers for tropical disease research, 1993. *Death and disease in the majority world: Keys to recognition and prevention*. Report of the second annual meeting. Bethesda: USPHS.

Plomin, R., et al., 1994. The genetic basis of complex human behaviors. *Science* 264:1733.

Plotkin, B., et al., 1997. Designing an international policy and legal framework for the control of emerging diseases: First steps. *Emerging Infectious Diseases* 3:1.

Post, R., 1997. Molecular biology of behavior: Targets for therapeutics. *Arch Gen Psych* 54:607.

Proctor, R., 1995. No time for heroes: Basic cancer research gets all the glory, but known preventive measures could save more lives. *The Sciences* March:20.

Rabinovich, N., et al., 1994. Vaccine technologies: View to the future. *Science* 265:1401.

Reiser, S., 1993. The era of the patient: Using the experience of illness in shaping the missions of health care. *J Amer Med Assn* 269:1012.

Shapiro, H., 1997. Ethical and policy issues of human cloning. *Science* 277:195.

Strauss, E., and S. Falkow, 1997. Microbial pathogenesis: Genomics and beyond. *Science* 276:707.

Summary, 1994. Addressing emerging infectious disease threats: A prevention strategy for the United States. *Morbidity and Mortality Weekly Reports* 43:RR5–1.

Tacket, C., et al., 1998. Immunogenicity in humans of a recombinant bacterial antigen delivered in a transgenic potato. *Nature Medicine* 4:607.

Whalen, R., 1996. DNA vaccines for emerging infectious diseases: What if? *Emerging Infectious Diseases* 2:168.

Williams, G., and R. Nesse, 1996. The dawn of Darwinian medicine. *Quarterly Review of Biology* 66:1–22.

Index